# STRATEGIES FOR MANAGING OZONE-DEPLETING REFRIGERANTS
## Confronting the Future

by

Katharine B. Miller          Charles W. Purcell
Jennifer M. Matchett          Marjut H. Turner

**BATTELLE PRESS**
Columbus • Richland

Library of Congress Cataloging-in-Publication Data

Strategies for managing ozone-depleting refrigerants : confronting the
    future / by Katharine B. Miller . . . [et al.].
            p.        cm.
        Includes bibliographical references and index.
        ISBN 0-935470-84-0 : $34.95
        1. Chlorofluorocarbons—Environmental aspects.    2. Refrigerations and
    refrigerant equipment—Cost effectiveness.    3. Product life cycle.
    4. Air quality management.        I. Miller, Katharine B., 1963–   .
    TD887.C47S77    1995                                                94-32043
    621.5′64—dc20                                                       CIP

Printed in the United States of America

Cover photo by Dan Forer, Forer, Inc.

Additional copies may be ordered through:
Battelle Press
505 King Avenue
Columbus, Ohio 43201, USA
(614) 424-6393 or 1-800-451-3543

# CONTENTS

# ACKNOWLEDGEMENTS

In any endeavor of this magnitude, there are always a number of people "behind the scenes" without whose support this book would not have been possible. The authors wish to acknowledge the significant technical contribution and assistance provided by the following staff from Battelle Pacific Northwest Laboratories: Jim Dirks, Graham Parker, Steve Weakley, Dave Hunt, Dave Conover, Jim Donaghue, and Janice Longstreth. The authors are also grateful for the support of Kelly O'Brien our summer intern who managed to keep up with whatever we threw at her, and Don Hanley who pulled the work of four people together. We would also like to recognize the representatives from York, Trane, Carrier, DuPont and ICI with whom we worked to get information on products and services. Special thanks goes to Don Anderson of Performance Associates; Kent Anderson of International Institute of Ammonia Refrigeration; Harry Gordon of Burt, Hill, Kosar, Rittelmann Associates; Tom Edwards of Trane; Steve Lauer of York; James Parsnow of Carrier; and Mike Weise of Trane. We would also like to acknowledge Mr. Theodore Koss with whom we have worked on ozone-depleting substance phaseout issues during the last several years.

KATHARINE MILLER
CHARLES PURCELL
JENNIFER MATCHETT
MARJUT TURNER
*Washington, DC 1994*

# FOREWORD

The stratospheric ozone-layer protects the earth against harmful ultraviolet radiation. The *Montreal Protocol on Substances that Deplete the Ozone-Layer* and the 1990 Clean Air Act phaseout the production of most ozone-depleting refrigerants, including chlorofluorocarbon (CFC) refrigerants, by the end of 1995. Hydrochlorofluorcarbon (HCFC) refrigerants will be controlled by 1996, reduced by 90% by 2015, and phased out no later than 2030. The world has set these demanding schedules because the ozone layer is already depleted to unacceptable levels, because depletion will worsen even after production is halted, and because it will be many decades before the ozone layer will recover.

These CFC and HCFC refrigerants have been the mainstay for refrigeration and air conditioning applications since their development in the late 1920s. The equipment that is affected by this phaseout represents a significant capital investment. For this reason, careful planning and analysis is required to ensure a timely, and cost-effective, transition to non-ozone depleting refrigerants.

*Strategies for Managing Ozone-depleting Refrigerants: Confronting the Future* provides technical information and a strategic business approach to managing ozone depleting refrigerants. It includes essential information on the availability of alternatives to ozone-depleting refrigerants, critical data to make an informed choice of equipment retrofit or replacement, and it explains how to arrive at a cost-effective solution. This book will be useful to environmental managers, building/facility mangers, supervisors, and anyone else with an interest in refrigerant management.

Katharine Miller and Charles Purcell have worked extensively with the Environmental Protection Agency, Department of Defense, Department of Energy, United Nations Environmental Programme, and the North Atlantic Treaty Organization (NATO) in promoting cost-effective alternatives and substitutes to ozone-depleting substances. They have been instrumental in the development of management plans for both large and small organizations (including the U.S. Department of Energy). Their considerable experience and wealth of information is presented in this new book.

Time is running out. Prompt action is needed to manage the existing supply of refrigerants and to retrofit and replace equipment to use the new alternatives and substitutes.

—DR. STEPHEN O. ANDERSEN
*Co-Chair Technology and Economic Assessment Panel*
*United Nations Environment Programme*

# INTRODUCTION

Today chlorofluorocarbon (CFC) and hydrochlorofluorocarbon (HCFC) refrigerants are used extensively in refrigeration and air conditioning applications around the world. Since their development in the 1920's, they have been hailed as safe, economical, and reliable refrigerants for use in numerous commercial and industrial applications. As early as the 1970's, however, scientists have been concerned that the same substances that we have found to be so safe in applications from refrigeration to fire suppression may be causing significant damage to the stratospheric ozone layer. As a result, an international decision was reached to begin to eliminate these substances.

## WHO SHOULD READ THIS BOOK?

This book is written to assist anyone who must develop strategies and approaches for dealing with the increasing price and scarcity of CFC refrigerants. This includes the dairy farmer trying to keep his milk chilled, the building manager of a small facility needing comfort cooling for his employees, and the manager of a large multi-facility installation with many different pieces of CFC dependent equipment that includes comfort cooling, refrigeration applications and process cooling.

The end of CFC production, and the attendant opportunity for equipment and refrigerant changes, presents a chance to increase energy efficiency throughout your facility. While many are viewing the CFC refrigerant production phaseout with trepidation, it can also be approached as an opportunity. The extent to which the end of CFC production will affect your operation is largely dependent on how well you have planned for the transition to alternative chemicals and technologies.

## HOW TO USE THIS BOOK

This book provides the necessary information to develop both short-term and long-term approaches for managing your existing ozone-depleting refrigerants and making plans for transition to alternatives. It should be used to develop an understanding of the range of alternatives and the complexities of the issues facing both technicians and managers. It offers an approach to prepare facilities and businesses for the future by ensuring timely, cost effective, responsible decisions are made to manage CFC based systems.

Since the decision to phase out CFCs and HCFCs in 1987, users of these compounds have been inundated with information on the issues surrounding the phaseout and actions that users should take to prepare for a CFC-free future. The sheer volume of the issues that users are being asked to consider can make the management of the phaseout seem an immense task. This book is designed to provide a strategy to managing CFC, and eventually HCFC, refrigerant phaseout using an approach that takes into account issues such as refrigerant availability, investment in refrigeration and air conditioning equipment, and energy efficiency.

It should be emphasized that there is no single "right way" to approach refrigerant management. The four sections are designed to provide background, answer questions, provide guidance, present various choices, and outline economic issues to help you ensure that the phaseout impacts your operations as little as possible. The most important requirement for users of CFCs is that they have some sort of plan that anticipates the production phaseout. That is the way to confront the future.

# STRATEGIES
# FOR
# MANAGING
# OZONE-DEPLETING
# REFRIGERANTS

## Confronting the Future

# Section I

## SETTING THE STAGE

# CHAPTER 1

# SCIENCE AND REGULATION

Since their development in the 1920's, the use of CFCs as refrigerants in vapor compression systems has become almost universal. These compounds have proven themselves to be not only highly stable, but also nonflammable and nontoxic, while having favorable cooling properties. In short, they have been praised as the ideal refrigerants. So why are we getting rid of them?

As most users of CFCs and HCFCs are aware, these compounds have been identified as environmentally harmful because they are capable of destroying stratospheric ozone. Since the reauthorization of the U.S. Clean Air Act in 1990, the U.S. Environmental Protection Agency (EPA) has been aggressively implementing regulations to limit the production and impact of CFCs, HCFCs, and other designated ozone-depleters. In fact, new regulations concerning CFCs have been published so quickly that many CFC users have found it difficult to keep track of, let alone read, them. The strict timelines established for phasing out the production of these substances have left some users wondering what the real issues behind ozone-depletion are. This chapter provides a brief and simplified overview of the ozone-depletion problem, the role played by CFCs and HCFCs in stratospheric ozone-destruction, and the potential adverse human health and environmental consequences that depleted stratospheric ozone can cause.

## WHAT IS THE STRATOSPHERIC OZONE LAYER?

The stratosphere is part of the earth's upper atmosphere, extending from approximately 8 to 30 miles above the earth's surface (Figure 1). The ozone found in the stratosphere is formed continuously as the result of reactions between oxygen molecules ($O_2$) and solar radiation. This radiation breaks the oxygen molecules into two oxygen atoms (O). These single atoms are very reactive and usually combine quickly to form another oxygen molecule ($O_2$). In the presence of solar radiation, however, the single oxygen atoms will also occasionally combine with an oxygen molecule (Figure 2) to form an ozone molecule ($O_3$).

$O_2$ + UV radiation from sunlight = O + O

$O_2$ + O = $O_3$ (ozone)

Ozone is constantly being broken apart by the solar radiation that it absorbs. This does not result in a net loss of ozone, however, because the process of ozone destruction produces another oxygen atom (O), which can then react with an oxygen molecule ($O_2$) to form more ozone ($O_3$).

There are several chemicals in the atmosphere that can react with and break apart ozone molecules. One of the more reactive chemicals is chlorine, which is a component of both CFCs and HCFCs and is also naturally emitted by the oceans. In 1974, the first connection between halocarbons, which include both CFCs and HCFCs, and ozone-destruction was established[1].

---

*Other halocarbons that scientists believe contribute to stratospheric ozone depletion are halons, carbon tetrachloride ($CCl_4$), methyl chloroform (1,1,1-trichloroethane), and methyl bromide.

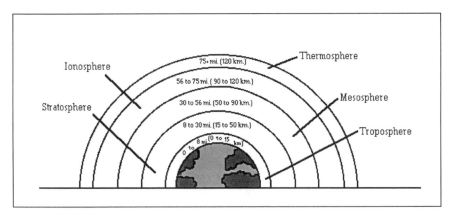

*FIGURE 1.*
*Diagram of the earth's atmospheric layers.*

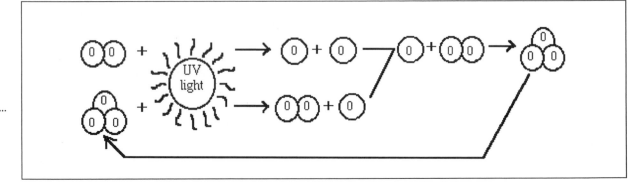

FIGURE 2. *Natural formation and destruction of ozone.*

Since 1978, the National Aeronautics and Space Administration (NASA) has been using a satellite-based instrument called the "Total Ozone Mapping Spectrometer" to measure the ozone layer and the amount of solar radiation it absorbs. Using this and other measurement techniques, scientists have detected a decrease in the amount of ozone over the Antarctic each spring. In 1991, the first evidence was presented of substantial seasonal ozone loss during both the spring and summer in the northern and southern hemispheres[1]. The net stratospheric ozone decrease has been attributed to the increased use and emission of halocarbons since they were first produced in the late 1920s. The reason CFCs, and to a lesser extent HCFCs, are theorized to contribute to ozone destruction is largely a result of their chemical stability. Many of the substances emitted at the earth's surface, both natural and man-made, break down in the lower parts of the atmosphere when they come into contact with water vapor. They then return to the earth's surface in dew or rain. Because CFC molecules are not very reactive, they do not break down in the presence of water vapor. For this reason, they are able to remain intact while being carried into the earth's stratosphere by air currents.

Once they reach the stratosphere, these molecules are impacted by the same solar radiation that forms ozone. The chemical bonds that hold the CFC molecules together are broken, releasing single chlorine ions (Cl-). When this free and highly reactive chlorine ion comes into contact with the relatively unstable ozone ($O_3$) molecules, it breaks the ozone molecule apart form-ing molecular oxygen ($O_2$) and chlorine monoxide (ClO). However, chlorine monoxide is itself unstable and will readily break down when it comes in contact with another ozone molecule. Figure 3 illustrates these reactions between chlorine and ozone. In this manner, the chlorine released from a single CFC can react with and destroy tens of thousands of ozone molecules.

Researchers believe it takes CFC molecules approximately 7 to 10 years to reach the upper stratosphere. With a half-life of 50 to 100 years[2], this means that CFCs that were released over the last decade are only now beginning to affect the ozone layer. This also means that control measures put in place today will not likely have a measurable impact on stratospheric ozone levels until some time in the future.

HCFCs are less likely to react with stratospheric ozone than are the CFCs because they contain hydrogen. The addition of a hydrogen atom makes the HCFCs less stable than CFCs and more susceptible to reaction with other naturally occurring chemicals in the stratosphere. In the atmosphere, sunlight breaks down water vapor ($H_2O$) and produces hydrogen (H+) and hydroxyl (OH-) ions. The hydrogen on the HCFC molecule reacts with the hydroxyl ion to form water, and the remainder of the HCFC molecule, in turn, forms non-reactive products (such as carbon dioxide), or acids (such as hydrochloric acid), that are removed from the atmosphere through other processes. The tendency to react with hydroxyl ions reduces the probability that HCFCs will reach the upper stratosphere to react with ozone.

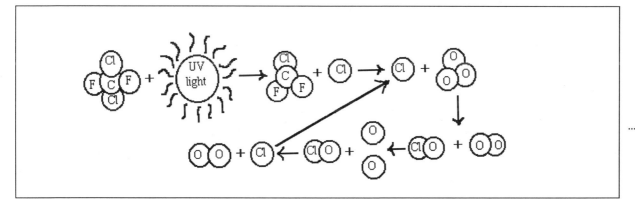

FIGURE 3. CFC induced destruction of ozone.

## OZONE-DEPLETION POTENTIAL

Scientists have categorized halocarbon molecules according to the rate at which they can react with, and destroy, ozone. The ozone-depletion potential (ODP) of a compound is determined relative to the effect of CFC-11. Or in other words, the ODP of CFC-11 is set to 1 and all other compounds are compared to CFC-11 when their ODPs are calculated. The more likely it is that the compound will remain unbroken from the time it is emitted to its arrival in the stratosphere, the higher the compound's ODP. As discussed above, one of the major factors that determines whether a compound will remain intact as it travels to the stratosphere is its ability to react and/or be degraded by other compounds. Another primary factor is the length of time that the reactive portion of the molecule remains in the stratosphere to react with ozone before becoming joined with other chemicals into a non-reactive product. This time period is called the "atmospheric life" and is also important in determining the compound's global warming potential.

## WHAT ABOUT GLOBAL WARMING?

There is a lot of confusion over the relationship between ozone-depletion and global warming. In fact, they are two different phenomena for which the same chemicals can play a part. Global warming, or the "greenhouse effect," is an increase in atmospheric temperatures resulting from the trapping of heat from the sun. In order for life to exist in earth, there needs to be a greenhouse effect. The natural greenhouse effect ensures that thermal energy from the sun is retained in the atmosphere rather than being reflected back out to space. There are five primary gases that accomplish this natural greenhouse effect. They are: carbon dioxide, methane, ground level ozone (smog), and water vapor. Increases in these primary gases have occurred over the last century as a result of human activity and development. This activity has also introduced another set of gases that have the ability to trap heat: the CFCs and HCFCs. As the amount of these gases increases, so the theory goes, so does amount of heat trapped in the atmosphere. The result is an overall increase in atmospheric temperatures.

Scientists have developed an index called the Global Warming Potential (GWP) to help them identify a chemical's ability to contribute to the trapping of solar radiation. The GWP is a relative index based on the global warming potential of carbon dioxide. The GWP is calculated by the amount of gas remaining in the atmosphere and its ability to absorb energy (heat) over a 100 year time-frame. The longer a chemical is able to remain in the atmosphere, the higher the GWP generally is. The contribution of CFCs and HCFCs to global warming is still being studied. Scientists are uncertain whether depletion of the stratospheric ozone contributes to or hinders global warming, and therefore whether the net effect of CFCs and HCFCs is positive or negative in terms of global warming.

Some of the alternatives that are being identified as replacements for CFCs and HCFCs also have global warming potentials. No one would suggest that a decision to replace an ozone-depleting refrigerant be made on the basis of global warming potential alone. There are many who argue that the energy efficiency benefits of certain alternative chemicals cancel out the impacts that these alternatives have in global warming. Whatever the final decisions on halocarbons and their substitutes turns out to be, it is important that users of CFC and HCFC refrigerants be aware of the increased emphasis being placed on the global warming issue both nationally and internationally. For example, the United Nations Environment Programme, which championed the controls for CFCs and HCFCs, has taken on this issue of global warming as its top priority for the foreseeable future[3]. This seems to suggest that compounds seen to contribute to global warming will find themselves controlled in the future.

## POSSIBLE HUMAN AND ENVIRONMENTAL IMPACTS OF STRATOSPHERIC OZONE LOSS

The decrease in stratospheric ozone levels has increased scientific concern about the environmental and health impacts such a reduction would have. As the stratospheric ozone layer is depleted, the amount of harmful ultraviolet radiation (UV-B)* that can reach the earth's surface increases. Increased exposure to UV-B radiation

---

*Ultra violet radiation is the radiation that is just beyond the violet spectrum. The harmful part of the band is the 280-320nm range and is called UV-B.

can affect human health, animals, and aquatic ecosystems.

In humans, the recognized harmful effects of exposure to UV-B radiation include suppression of the immune system; eye disorders, including cataracts and nearsightedness; and skin cancer, i.e., melanoma and non-melanoma. The United Nations Environment Programme (UNEP) estimates that a 10% decrease in stratospheric ozone could result in a 26% increase in non-melanoma skin cancer, which translates to an increase of over 300,000 cases per year worldwide. Furthermore, a 10% decrease in stratospheric ozone could result in 1.6 to 1.8 million additional cases of cataracts per year worldwide[4].

Relatively little research has been conducted on the effects of increased UV-B radiation on animals that live on land. Most researchers believe that these animals are protected from the effects of UV-B radiation by their fur, feathers, and scales. Additionally, many nonmammals lay eggs that are opaque to UV and are also usually shielded by a nesting parent. Hence, most concern for animals has been focused on increases in eye and skin disorders because their eyes and noses are not covered or shielded from the sun's direct rays. However, researchers do think UV-B radiation may damage amphibian eggs, which are not opaque or shielded from the sun's rays.

Research on the effects of UV-B radiation on aquatic ecosystems has primarily focused on phytoplankton. Phytoplankton are microscopic marine plants that live in the region of the ocean that sunlight penetrates (the photic zone). They are a primary source of food for many aquatic species. Phytoplankton also absorb carbon dioxide and create oxygen through photosynthesis just as land plants do, and it has been estimated that through their photosynthesis, phytoplankton annually convert approximately 104 billion tons of carbon into organic material — nearly 4 billion tons more than all terrestrial ecosystems combined[5]. Research has indicated that increased UV radiation can inhibit the growth of some types of phytoplankton.

## THE MONTREAL PROTOCOL ON SUBSTANCES THAT DEPLETE THE OZONE LAYER

Concern over the potential human and environmental health impacts of ozone-depletion led to the development of an international agreement in 1987 to reduce the production of halocarbons. To date, more than 131 nations have signed the *Montreal Protocol on Substances that Deplete the Ozone Layer*.

The original agreement called for a 50% reduction in production of certain ozone-depleting substances from their 1986 levels. Since the original Protocol was developed, increased evidence that ozone-depletion is occurring more rapidly than originally anticipated has twice resulted in the acceleration of the schedules for phasing out production of halocarbons. The 1990 London Amendments called for a total ban on the production of CFCs by January 1, 2000. The 1992 Copenhagen Amendments accelerated this timetable to complete the phaseout by December 31, 1995. The schedule for HCFCs was also accelerated to complete phaseout by 2030.

## THE CLEAN AIR ACT AMENDMENTS OF 1990

The United States codified the requirements of the *Montreal Protocol* in Title VI ("Stratospheric Ozone Protection") of the Clean Air Act Amendments of 1990. In addition to establishing a procedure for the production phaseout, the Clean Air Act also includes a provision to limit emissions of CFCs and HCFCs during routine maintenance and operation of refrigeration and air conditioning equipment. The Clean Air Act requires proper labelling of containers of CFCs and HCFCs, certification of technicians who service refrigeration and air conditioning equipment, certification of refrigerant recycling equipment, and mandatory practices for equipment disposal. The Clean Air Act also establishes a program to evaluate CFC and HCFC substitutes to ensure that they are not harmful to human health and the environment. Figure 4 shows the production phaseout schedules for refrigerant CFCs and HCFCs.

| Production Phaseout Date (January 1) | Refrigerant |
|---|---|
| 1996 | all CFCs |
| 2020 | HCFC-22 |
| 2030 | all other HCFCs |

FIGURE 4. Production phaseout schedules for refrigerant CFCs and HCFCs.

## REFERENCES

[1]*Clean Air Report* (CAR). 1992. "Antarctic Ozone Depletion Worse Than Ever, U.S. Researchers Say." December 17, 1992. Inside Washington Publishers, Washington, D.C.

[2]Salvato, J. *Environmental Engineering and Sanitation, 4th Edition.* Wiley/Interscience, John Wiley and sons; NY, NY, 1994, p. 774.

[3]United Nations Environment Programme (UNEP), 1989; *Action on Ozone.* United Nation's Environment Programme, Nairobi, Kenya. August 29, 1989.

[4]United Nations Environment Programme (UNEP), 1991; *Environmental Effects of Ozone Depletion: 1991 Update.* UNEP Environmental Effects Panel Report; Nairobi, Kenya.

[5]United Nations Environment Programme (UNEP), 1991; *Environmental Effects of Ozone Depletion: 1991 Update.* UNEP Environmental Effects Panel Report; Nairobi, Kenya.

# CHAPTER 2

# REFRIGERANTS AND EQUIPMENT

## INTRODUCTION

This chapter provides an overview of general principles of refrigeration and refrigeration equipment. This information is provided as background for readers with an interest in the types and operations of refrigeration and air conditioning systems. The types of equipment discussed here are those that are currently commercially available and have been used in various applications in the past. In reviewing your options for eventually phasing out your use of ozone-depleting refrigerants, it may be valuable to review the other types of systems that are currently in use.

## REFRIGERATION

Refrigerants are the working fluids in refrigeration, air conditioning, and heat pump applications. Refrigerants are generally referred to by a numbering system developed into a standard by the American Society of Heating, Refrigeration and Air-Conditioning Engineers (ASHRAE) and accepted by the International Organization for Standardization. Under this system, a two-to-four-digit number is used to designate the refrigerant's chemical formula. Starting from the right-hand side, the first digit is the number of fluorine atoms in the molecule. The second digit is the number of hydrogen atoms plus 1. The third digit from the right is the number of carbon atoms minus 1. This number is omitted if it is a 0 (i.e., a single carbon). The fourth digit is used to indicate whether a carbon-carbon double bond is present. If such a bond exists, the fourth digit is a 1. If the carbon-carbon bond does not exist (i.e., a saturated compound), the fourth digit is omitted. An example of how this numbering process works for CFC-11 and CFC-123 is shown in Figure 1. Numbers are also arbitrarily assigned to refrigerant blends (400 series), azeotropes (500 series), miscellaneous organic compounds (600 series), and inorganic compounds (700 series).

The numbers described above are generally preceded by the prefix letter "R," designating a refrigerant. The ASHRAE standard allows the use of prefixes that indicate the chemical composition of the refrigerant (such as CFC, HFC, HCFC, and HC) to precede the numbers in the place of R. A CFC indicates a molecule composed of chlorine, fluorine, and carbon. An HCFC designates a molecule composed of hydrogen, chlorine, fluorine, and carbon. An HFC designates a molecule composed of hydrogen, fluorine, and carbon. An HC designates a hydrocarbon. These designations are useful in identifying which refrigerants are chlorinated compounds and which are not. Trade names such as Freon are also often used in place of the R or CFC, although this practice is not approved by ASHRAE.

There are several criteria that make a chemical a good candidate to be used as a refrigerant. These include chemical stability, nonflammability, low toxicity, favorable thermophysical properties, and compatibility with a number of materials used in refrigeration and air conditioning equipment. The halocarbons have dominated as refrigerants because of their ability to meet these sometimes conflicting refrigerant requirements. The most commonly used halocarbon refrigerants are R-11, R-12, R-13, R-113, R-114, R-500, R-502 and R-22. R-500 is a blend of R-12 and R-152a. R-502 is a blend containing R-22 and R-115.

## STABILITY

Stability is a primary requirement for a good refrigerant. The refrigerant must be capable of working in the system without decomposition in

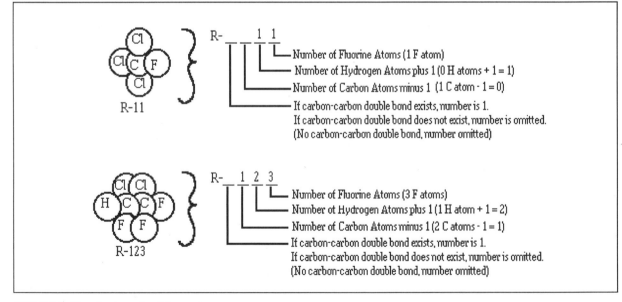

FIGURE 1. Numbering of refrigerants.

order to maintain its thermodynamic properties and achieve the desired cooling level. Additionally, because many cooling systems use configurations that locate the motor and compressor in the same pressure vessel, the refrigerant is in direct contact with the motor. These sealed systems require the refrigerant to remain in service for the life of the equipment. The refrigerants are subjected to very rigorous operating conditions, and stability is vital to ensure effective, efficient operation. Refrigerant blends (such as R-500) must be stable enough to maintain their relative composition and not break down into the individual constituents over time.

## SAFETY AND HEALTH

In the United States, there are numerous safety codes that limit the applications in which flammable or toxic refrigerants can be used. The most stringent safety standards have been established for cooling occupied spaces in residential or commercial structures. Halocarbons have proven themselves to be safe refrigerants due to their low toxicity and flammability. In fact, the development of the halocarbons in the 1920s was based on an assumption that domestic refrigeration required safer refrigerants than the ammonia and carbon dioxide in use at that time. It must be stated, however, that the tests for toxicity and flammability have been changed significantly since the halocarbons were first tested. Thus, some of the refrigerant alternatives are required to meet more severe criteria in the areas of flammability and toxicity than the halocarbons to which they are being compared. In addition, the safety of the refrigerant cannot be separated from the overall safety of the equipment in which it is used. Due to advances in design, refrigeration equipment today is significantly safer than earlier models, and the range of refrigerants that can be used has increased. For example, some European countries are evaluating the use of propane as a refrigerant in household refrigerators. The argument for this is the increased safety of the refrigerator structure, including refrigerant containment ability.

An additional criterion under the safety and health category is the impact that the refrigerant has on the environment. The stability of the halocarbons has contributed to their overall safety in terms of flammability. However, it is this same stability that has allowed them to remain in the atmosphere over long periods of time with potential damaging consequences for the environment. As with the relationship between safety and equipment, there is also a relationship between environmental impact and containment. Refrigerants that are tightly contained and responsibly used will have minimal impacts on the environment even if their ozone-depletion potential is very high.

## THERMOPHYSICAL AND THERMODYNAMIC PROPERTIES

Both absorption and vapor-compression refrigeration processes take advantage of the change in state, or phase, of refrigerants brought about by the application of energy or heat. The thermophysical properties of a compound determine whether that compound is suitable for use as a refrigerant, as well as indicating the potential efficiency and economy of the refrigeration cycle. Selecting a refrigerant often involves tradeoffs between conflicting desirable thermodynamic qualities, such as the efficiency of the system and its cooling capacity. In general, the efficiency tends to increase as the critical temperature of the refrigerant increases. The capacity, however, tends to decrease in relation to increasing critical temperature. Since efficiency determines the operating cost while capacity determines the capital cost of the equipment, this tradeoff has implications for the economics of system operation.

Refrigerants are categorized as "low-," "medium-," or "high-pressure" refrigerants. These terms describe the relative pressure and temperature at which the refrigerant vaporizes. Existing refrigeration and air conditioning systems are optimized for the refrigerants that they currently use. The critical pressure of the refrigerant will determine the pressure range of the compressor.

## SOLUBILITY IN LUBRICANT

In addition to lubricating equipment components, lubricants in refrigerant systems also act as coolants to remove heat from the bearings and to transfer heat from the crankcase to the compressor exterior. In hermetically sealed (or airtight) systems, the refrigerant vapor carries some lubricant with it into the condenser and evaporator. The lubricant and refrigerant must

11

remain mixed to provide good heat transfer in the evaporator. If the lubricant and refrigerant become separated and the lubricant adheres to the condenser or evaporator tubes, it can adversely affect the heat-transfer characteristics of the component. Lubricants in hermetic systems must last the lifetime of the compressor, which is generally between 25 to 30 years. The lubricant-refrigerant mix should remain stable over a long period of time and long operating hours.

## MATERIALS COMPATIBILITY

Ensuring compatibility between the refrigerant and other materials used in refrigeration and air conditioning equipment is especially important when considering the use of a refrigerant other than the one for which the system was designed. Some alternatives to CFCs and HCFCs have potential compatibility concerns with lubricating oils, seals, hoses, and motor insulations, as well as the metals used to fabricate heat exchangers and compressors. In addition, some refrigerants, such as ammonia, are incompatible with certain materials, which can require the use of non-standard or specialized materials.

## EQUIPMENT

There are several approaches for achieving air-conditioning and commercial and industrial refrigeration. The most common processes are

- Vapor Compression
- Absorption
- Evaporative Cooling
- Desiccant Cooling.

Equipment using the vapor compression cycle is the most common type of equipment for industrial and commercial refrigeration and air conditioning applications. This process has become dominant because of its superior cooling efficiency and relative simplicity. A large portion of the vapor compression systems in operation today utilize halocarbon refrigerants, thus; it is these systems that will be most affected by the CFC and HCFC production phaseout.

## VAPOR COMPRESSION

In the basic vapor compression process (Figure 2), a liquid refrigerant is "evaporated" by absorbing or removing heat from the air or space to be cooled. The gaseous refrigerant is compressed to allow the heat in the refrigerant

12

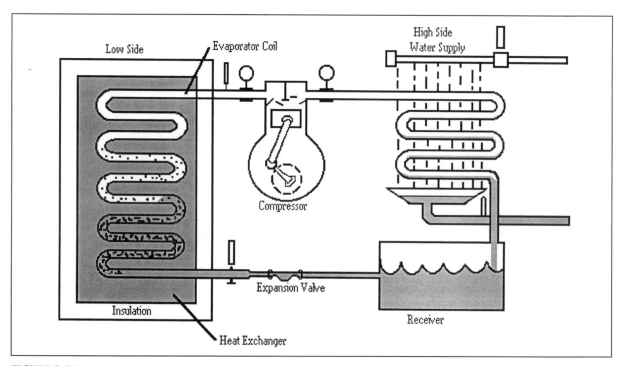

*FIGURE 2. Vapor compression system.* (Courtesy of International Institute of Ammonia Refrigeration)

to be ejected into a higher temperature sink - usually the ambient air outside the cooled space. The compressed refrigerant is passed through an expansion device where it cools and returns to its liquid state. The process is then repeated.

A vapor compression system has both a high- and a low-pressure side. These are relative terms. The low-pressure side is in the heat exchanger before the refrigerant is compressed. The high-pressure side is after the compressor. Depending on the type of refrigerant used, the low pressure side operates at pressures below ambient or outside pressures. Refrigerant leaks in such "low-pressure" equipment result in air and moisture leaking into the equipment, rather than refrigerant leaking out. The air and/or moisture must then be purged from the system which generally results in refrigerant loss. High-pressure equipment uses refrigerants that are liquids at pressures above ambient or outside pressures. Refrigerant leaks in this equipment result in the loss of refrigerant to the atmosphere.

In the vapor compression cycle, the refrigerant is condensed when it passes through the expansion device and returns to its liquid state (see Figure 2). Methods for cooling the refrigerant usually involve the use of air at ambient temperature or cooling water from some source. The temperature of these cooling media can affect the extent to which the refrigerant is cooled. Equipment performance, in terms of both energy efficiency and capacity, will also be affected by the temperature of the cooling medium in the condenser. Some systems have required temperatures for the cooling medium that must be met for the equipment to operate effectively. Failure to meet these requirements may result in loss of efficiency or damage to the equipment. In contrast, it may also be possible to increase the energy efficiency of certain equipment by decreasing the temperature of the cooling medium.

The different types of compressors that will be discussed below generally have the option for using either water-cooled or air-cooled condensers. Figure 3 shows the differences in coefficient of performance, or efficiency, for each type of condenser. Generally speaking, water-cooled condensers provide the best coefficient of performance.

| COMPRESSOR TYPE | kW/ton | COPs |
|---|---|---|
| Reciprocating Compressor | | |
|   Air | 1.2 | 2.9 |
|   Water | 1 | 3.5 |
| Screw Compressor | | |
|   Air | 1.05 - 1.1 | 3.4 - 3.2 |
|   Water | 0.65 - 0.75 | 5.4 - 4.7 |

FIGURE 3. Comparison of condenser cooling mediums on performance for two compressor types (ARI standard conditions).

## Cooling Towers

The cooling process produces heat that has been removed from the air that is cooled and which must be removed or dissipated from the system. There are several methods for accomplishing this. For a liquid-cooled system, a common practice is to use a cooling tower. In an air-conditioning application, cooling water is circulated through the condenser's heat exchanger to a cooling tower. In industrial process applications, the cooling tower water is circulated through the process heat exchanger then to the cooling tower. The cooling tower uses an evaporation process to reject the heat that the water has gained while traveling through the condenser to the ambient air. In the cooling tower, water is circulated through various systems of sprays nozzles, splash bars or louvers to enhance evaporation. Cooling towers tend to be a high-maintenance component of a cooling system.

## Methods of Transferring Cooling

There are two methods for providing cooling to a space, product, or process: the direct and indirect methods. In the direct method (see Figure 4), the refrigerant air is moved through a space and across a heat exchanger. This air gives up its heat to the refrigerant fluid in the heat exchanger and the "cooled" air moves into the space thereby cooling the space. In the indirect method (see Figure 5), the refrigerant is used to cool a secondary fluid such as water or a brine, which is then circulated to a heat exchanger to provide the cooling to the space or process. Both of these methods are used extensively in

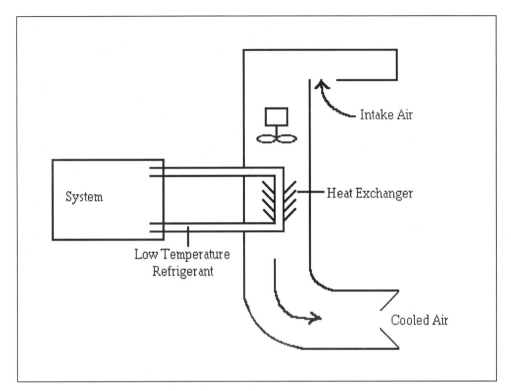

FIGURE 4. Direct cooling approach

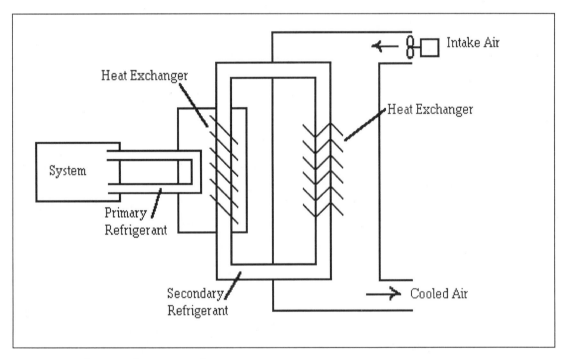

FIGURE 5. In-direct cooling approach..

the refrigeration and comfort cooling applications. In process-specific industrial applications variations of these techniques are applied.

## COMPRESSOR DRIVES

Vapor compression equipment is generally differentiated by the type of compressor that is used. There are two types of drives that can be used on different vapor compression systems. Because of their design, these drives are termed "open drive" and "closed drive." Each design has unique advantages and disadvantages.

### Open Drive

The "open drive" has the motor-driver exposed to the air and is connected to a drive shaft and, either directly or through a gear mechanism, to the impeller that serves as the compressor. The drive motor in the open system is not "hermetically" sealed but is exposed to the open air of the mechanical room. The only place where these systems will leak refrigerant is at the main drive shaft seal. The open system makes changing the motor-driver much easier than in a closed system; such a system is also less susceptible to material compatibility issues associated with substitute refrigerants. Systems using ammonia (R-717) are only designed in the open configuration because ammonia is incompatible with the copper wiring in the motor. When water and ammonia are present with copper, the copper disintegrates. In the open system, the mechanical room must have sufficient ventilation to be able to "cool" the motor.

### Closed Drive

In a "closed drive" system, the motor and the compressor are sealed within the same pressure vessel, with the motor shaft itself being an integral part of the compressor drive shaft. The motor is generally smaller because it is in direct contact with the refrigerant, allowing it to operate at a lower temperature and thus more efficiently. The compressor loses some of its efficiency because some of the cooling capacity is used to cool the motor. The motor must be compatible with the refrigerant and lubricant. Additionally, closed drive systems have more leak paths, and a motor change-out becomes more complex than with the open system.

### Centrifugal Compressors

The centrifugal compressor, sometimes called a "turbocompressor," is part of a family of machines that uses fans, propellers, and turbines. Refrigerant entering the compressor is spun by the compressor blades. The momentum imparted to the refrigerant by the spinning blades causes the refrigerant to be compressed against the sides of the compressor walls. Because of their ability to produce high pressure ratios and high volumetric capacities, centrifugal compressors are well suited to air-conditioning and refrigeration applications. Their volumetric capacity also allows them to have a relatively smaller physical size than reciprocating machines of equivalent capacity. Additionally, because of the wide range of pressure ratios (2 to 30), a very wide range of refrigerants can be used. Generally the refrigerants used are "low-pressure," (i.e., having a sub-atmospheric operating pressure). These low-pressure refrigerants include R-11 and R-114.

Centrifugal compressors, like the other compressor systems, can be driven by many different "prime movers." These include electric motors, turbines, or natural gas engines, ranging in power from about 30 to over 12,000 hp. The centrifugal compressors normally have cooling capacities ranging from 80 tons to 2000 tons (280 kW to 7000 kW) and up to 10,000 tons (35,000 kW) for specially designed systems assembled on-site.

### Reciprocating Compressors

Reciprocating compressors, also known as positive-displacement compressor systems, are considered high-pressure type systems. These compressors generally use the high-pressure refrigerants, such as R-12, R-22, R-500, R-502, or ammonia. Most of the reciprocating compressors use pistons driven directly through a piston rod connected to a crank shaft. The prime mover turns the crankshaft and drives the compressor pistons. Like the centrifugal compressor, the reciprocating compressor can be driven by both an open drive and a closed drive. The reciprocating compressor typically covers a capacity range of 10 tons to 120 tons (35 kW to 422 kW) and is very competitive in that range. The coefficient of performance and efficiency is generally lower for the reciprocating system than for other types of equipment.

## Screw Compressors

The screw chiller is a positive-displacement, high-pressure type machine and uses both open and closed drive systems. The screw compressor uses the meshing of the grooves in a large-toothed rotor to move the working fluid along into a compressed condition. The working fluids are the high-pressure refrigerants R-12, R-114, R-502, and the HCFC R-22, as well as ammonia. The modern systems with open drive cover the range from approximately 20 tons to 2000 tons (70 kW to 7000 kW). The "packaged" screw compressor that utilizes a closed or hermetically sealed drive system is used from the 40 ton up to the 850 ton (140 kW to 2975 kW) capacity range. While the screw compressors are somewhat more expensive to purchase (first cost), they have some advantages. These include smaller physical size for a given capacity than the centrifugal or the reciprocating compressors, and quieter and smoother operation. There are many new variations on the screw compressor design, including twin screws that enhance performance and energy efficiency.

## Scroll Compressors

The scroll compressor is used in residential and commercial air-conditioning and heat pump applications as well as in automotive air-conditioning. Like the screw and reciprocating machines, it uses a high-pressure refrigerant, such as R-12 and R-22. The technology employed in the scroll compressor is very similar to that used in the screw type compressor except that scroll chillers are generally used in the lower tonnage ranges. The general range of cooling capacity of this type of compressor is 1 ton up to 60 tons (3.5 kW to 210 kW). The scroll compressor is a rotary-type machine that is considered positive displacement. These compressors achieve the desired pressures through meshing two spiral-shaped scroll members machined to very close tolerances. The geometry of the scrolls is such that the volume at the entry point is greater than at the exit. As the refrigerant is moved through the scroll, the volume is progressively reduced and the refrigerant is compressed. At a selected point, the compressed refrigerant is discharged. The geometry of the scroll and the position of the discharge port give the scroll compressor different performance characteristics than those of either the centrifugal or reciprocating compressors.

## Condensing Units

These vapor-compression units are generally self-contained systems that are not true chillers. Rather, they are a combination of compressor and condenser requiring a separate evaporator and throttling valve located elsewhere — often in another room or building. These units use R-12, R-22, and R-502 as their refrigerants. Condensing units are typically employed in supermarket applications, though they can be used in other commercial, industrial, and residential applications as well. These units are designed to provide a specific cooling capacity at a given compressor horsepower and evaporator temperature. Condensing units may employ either an air-cooled or water-cooled condenser. The air-cooled units tend to be larger and more expensive and are not as efficient as water-cooled systems. However, the air-cooled units do not have the maintenance requirements of the water-cooled units, nor do they require the use of a cooling tower. Both air-cooled and water-cooled condensing units produce significant noise, which could be a disadvantage in some applications.

## AMMONIA VAPOR COMPRESSION

Ammonia is an alternative refrigerant that can be used in any of the compressors previously discussed. The only exception that its use requires an open drive because of ammonia's incompatibility with the copper windings used in motors. As a refrigerant, ammonia has several highly desirable qualities. These include good mixing capabilities with lubricants, very good heat transfer and thermochemical characteristics, low cost, easy leak detection because of the odor, and a long history of use throughout the world. The major drawbacks to ammonia systems are flammability and toxicity. These disadvantages can be mitigated in certain instances through engineering design, use of secondary coolants, and isolation of the portion of the system that contains the ammonia charge. Ammonia is the primary refrigerant for use in cold storage applications throughout the world.

## HYDROCARBON VAPOR COMPRESSION

The hydrocarbon refrigerants can also be used in the vapor compression cycle. These refrigerants are generally highly flammable and are therefore subject to strict regulations and codes regarding their use. The hydrocarbon refrigerants are currently used primarily in the chemical and petroleum industries, however; they are attracting considerable interest in Europe and other countries for various applications as the production of CFCs is ended. These countries generally have less stringent product liability codes that allow them to make use of these very effective refrigerants in applications that include domestic refrigerators.

## ABSORPTION REFRIGERATION

The absorption process has been used extensively in industrial process applications and for large tonnage chillers for Heating, Venting, and Air-Conditioning (HVAC) applications. The basic absorption refrigeration process (see Figure 6) uses a heat source to boil and vaporize a low-pressure refrigerant in an evaporator. The refrigerant vapor then moves to an absorber where it is absorbed into solution with another compound (absorbent) that has a high chemical affinity for the refrigerant. The mixed fluid is pumped through an air-cooled or water-cooled heat exchanger, where heat is removed. The mixture is then pumped to a generator where heat is used to drive the refrigerant from the absorbent. The concentrated absorbent is returned to the absorber, and the refrigerant vapor migrates to a condenser, where it is liquified by transferring its heat to the outside air or cooling tower. It then returns to the evaporator to begin the cycle again.

The primary refrigerants used in the absorption cycle are ammonia and water, or lithium bromide and water. The absorption chillers have a capacity range of approximately 50 tons to nearly 1700 tons (175 kW to 6000 kW). Small units using an ammonia-water pair are available in the 3-5 ton range.

A source of heat is required for the absorption cycle to continue and to provide the energy to separate the refrigerant from the absorbent in the generator vessel. This heat source can be from many sources, including

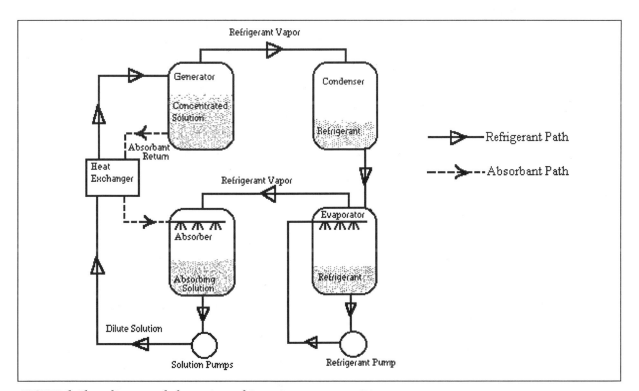

*FIGURE 6. Flow diagram of absorption refrigeration.* (Courtesty of Trane, Inc.)

17

- Direct fire gas or oil
- Indirect steam heat from boilers
- Waste process steam
- Hot process fluids
- Hot water solar, diesel, or gas engine-heated
- Cogeneration steam or hot water
- Centrifugal compressor steam turbine exhaust.

The ability of these systems to use waste heat has made them attractive cooling applications in some industrial situations where sources of waste heat are readily available.

## EVAPORATIVE COOLING

Evaporative cooling achieves refrigeration through the natural cooling process of water evaporating into the air. Evaporative cooling systems can be either direct or indirect. In the direct evaporative cooling process (see Figure 7), water is used as a refrigerant to absorb heat from the air in the space to be cooled. Air to be cooled is drawn across water-saturated pads by the suction action of the blower. The heat from the air causes evaporation of some of the water. This evaporation process lowers the temperature of the air, which is then blown into the space to be cooled. The amount of heat removed from the air is equal to the amount of heat absorbed by the water as heat of vaporiza-

tion. An indirect evaporative cooling process uses a mechanism (such as a desiccant) to dry the air, then cools the air via a heat exchanger. Traditionally, evaporative cooling systems were only used in hot, arid regions, but the indirect systems have helped broaden the geographic regions where evaporative systems can be effectively used.

Evaporative cooling can be an energy-efficient and cost-effective means of cooling in some situations. While the initial cost of evaporative cooling equipment may be higher than the typical CFC- or HCFC-based systems, the energy efficiency advantages can offset these initial costs. Because it relies on evaporation of water to cool the air, evaporative cooling does not work well in situations where the air has a high moisture content unless the air has been dried prior to entering the evaporative cooling process. Evaporative cooling can be used for comfort cooling in commercial buildings and industrial applications. In some cases retrofits may be difficult because the existing air handling ducts may be undersized to properly handle evaporative cooling systems.

## DESICCANT COOLING

Desiccant cooling is a modification of the evaporative cooling technology and, in many cases, is used in conjunction with evaporative

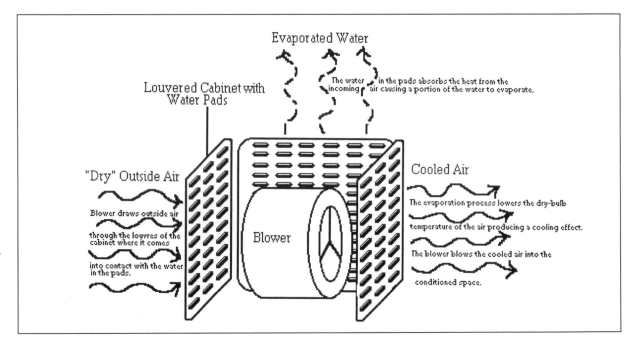

*FIGURE 7. Evaporative cooling.* (Brumbaugh, James E. Heating, Ventilating, and Air Conditioning Library, Volume III, 1987, pg. 369)

cooling systems. Desiccants are materials, such as titanium silicate, that attract and hold moisture from the surrounding air. They have traditionally been used to reduce moisture in air required for applications that have tight humidity controls. Identification of new desiccants are enabling the design of air conditioning systems.

Desiccant cooling systems (see Figure 8) first pull in hot air, usually from the return loop or low-cost sources such as exhausts or steam generated by gas-fired boilers. This hot, moist air is passed over a *dehumidification wheel*, which is usually a rotating, honeycomb-shaped wheel coated with a desiccant. The desiccant removes water and many common airborne pollutants from the incoming air. On the other side of the dehumidification wheel is hot air which "reactivates" the desiccant by picking up the moisture and ejecting it back to the outdoor air. After passing over the dehumidification wheel, the resulting hot, dry air is cooled and re-humidified using an evaporative process. The use of the desiccant system in conjunction with evaporative cooling extends the use of these systems to areas of high humidity where evaporative cooling alone may not be practical.

These systems tend to be physically quite large and, as such, have been used primarily in commercial applications. While desiccant cooling systems can be used in many situations, they are becoming increasingly popular as replacement systems in supermarkets. Desiccant systems remove moisture from the air, helping prevent frost and ice from forming on frozen foods without having to "overcool" the surrounding air to reduce the humidity.

Desiccant cooling systems have several benefits over standard vapor compression systems. Since desiccant systems dehumidify by absorbing moisture from the air rather than chilling air to condense moisture out, they do not have the wet cooling coils, drip pans, and humid ductwork generally associated with vapor compression systems. Eliminating these moist areas (which can act as breeding grounds for mold, mildew, fungi and bacteria) can help improve indoor air quality. These systems can also lead to improved indoor air quality since the desiccants,

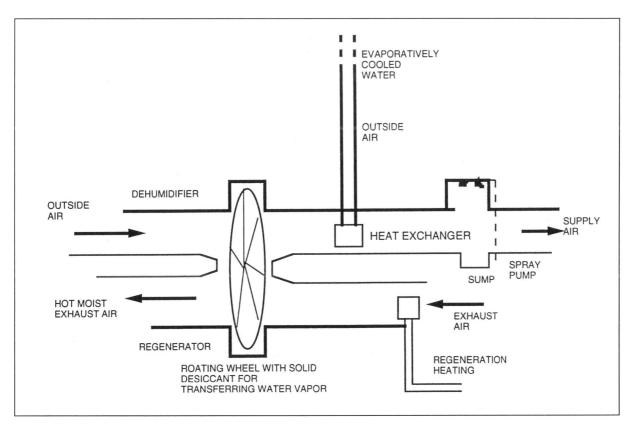

*FIGURE 8. Desiccant/evaporative cooling system without refrigerant.*

in addition to removing moisture, can remove up to 90 percent of airborne pollutants from the air[1]. Moreover, desiccant cooling systems typically use either low cost natural gas or waste heat as the primary energy source. Hence, there is a lower electricity requirement, which can result in energy savings for the user.

## CONCLUSIONS

Figure 9 shows the refrigerant equipment discussed in this chapter with the common refrigerants used. Significant changes are being made in both traditional and non-traditional air conditioning and cooling applications as a result of the phaseout of ozone-depleting refrigerants. Equipment is being improved in both performance and capacity. More cooling is being accomplished with smaller charges; more work is being accomplished with less energy; and the systems are becoming tighter with fewer leaks.

This chapter provided an overview of some of the most promising, currently available, technologies for providing air conditioning and refrigeration. Other promising technologies are also being developed that, should they prove viable, may be available for consideration in the next several years. While vapor compression technologies are likely to retain their dominance after the phaseout of ozone-depleting refrigerants, it is important to review other options that may be available as replacements for your current refrigeration and air conditioning applications.

## REFERENCE

[1]*The Air Conditioning, Heating, and Refrigeration News*. Business News Publishing Co: Troy, MI. "What desiccant-based cooling is, how it works." December 6, 1993. Vol. 190, No. 14. Page 3.

| *EQUIPMENT* | *REFRIGERANT* |
| --- | --- |
| Centrifugal Compressors | R-11, R-22, R-114 |
| Screw Compressors | R-12, R-22, R-114, R-502 |
| Scroll Compressors | R-12, R-22 |
| Reciprocating Chillers | R-12, R-22, R-500, R-502 |
| Absorption | Lithium bromide-water, Ammonia-water |
| Gas-Fired Chillers | Ammonia, R-12, R-22 |
| Heat Pumps | R-22 |
| Condensing Units | R-22 |
| Evaporative and Desiccant Cooling | Water, Salt spray |
| Ammonia Chillers | Ammonia |

*FIGURE 9. Equipment and refrigerants used.*

# CHAPTER 3

# REFRIGERANT Q AND A's

The previous chapters provided an overview of the reasons behind the production phaseout of halocarbons, and described the typical applications for halocarbon and non-halocarbon refrigerants. One of the most common misconceptions about the phaseout of CFCs and HCFCs is that there is no need for the equipment owner or operator to do anything. The assumption is that when the refrigerants are gone, others will be supplied to take their place that will not require any changes to be made to existing equipment. It is also assumed that because the phaseout is oriented toward refrigerant production, there are no requirements on refrigerant users.

These assumptions are not only wrong—they're dangerous. In fact, there are several requirements that refrigerant users must know about and comply with. Being fully informed is the best protection against disruptions in your operations—and the best way to avoid EPA citations and fines. In addition, the refrigerant manufacturers have been warning equipment owners not to assume that a "silver bullet" replacement refrigerant will come along that can be directly used in all existing systems. Retrofits of different magnitudes are required for existing equipment to use most of the common alternatives. This chapter answers some of the most common questions regarding the refrigerant phaseout, the applicable regulations, and how refrigerant users and equipment owners may be affected.

## Q: Am I required by law to stop using CFCs and HCFCs?

There is no requirement under either the Clean Air Act or the *Montreal Protocol* that users of CFCs and HCFCs discontinue use of these compounds. The Clean Air Act is targeted toward eliminating the production of CFCs and other ozone-depleting substances under the assumption that this will eventually eliminate the use of these compounds. Therefore, people who use these compounds in refrigeration and air-conditioning applications may continue to use them as long as they can find a source of supply.

In an unusual twist in its regulation development process, the Environmental Protection Agency imposed the first limitations on the use of HCFCs in the regulations accompanying the acceleration of the phaseout schedule on December 10, 1993. This regulation limits the use of certain HCFCs in applications other refrigeration. Users should be aware that HCFC-22, which is currently used in a number of applications and is a favorite replacement for some CFCs, is scheduled for production phaseout by 2020 (this is 10 years earlier than most other HCFCs). Although most HCFCs are an acceptable interim substitute, plans for their eventual phaseout should become part of any refrigerant management planning activity.

## Q: Can I still vent refrigerants while servicing, maintaining, and disposing of my equipment?

No. As of July 1, 1992, the intentional venting of refrigerant in the course of servicing, maintaining, repairing or disposing of appliances is prohibited by law. This prohibition

applies to both CFCs and HCFCs. There are three types of allowable releases:

- *"De minimus"* releases vented in the course of making good faith efforts to recycle or recover refrigerant

- Refrigerants emitted during the normal operation of equipment, such as from leaks (with some limitations) and mechanical purging

- Emissions of mixtures of nitrogen and HCFC-22 used as holding charges or as leak test gases

## Q: What does the EPA consider *"de minimus"* releases and "good faith" efforts to recover refrigerant?

The EPA has not set a numerical value to define *"de minimus"* releases. If you are properly using EPA-certified recovery equipment, any releases that occur *should* be minimal, and hence not considered venting. The key is that equipment must be certified and capable of meeting the required evacuation levels listed in Figure 1; and you must be using the equipment properly.

The EPA is defining "good faith efforts" to recover and recycle refrigerant as those efforts that are in compliance with the requirements established for recovery and recycling equipment and its operation. Basically, you need to be able to show that you are trying to comply with the requirements, either by having and using approved recovery equipment when maintenance or service is performed, or by contracting the servicing and maintenance of equipment to contractors who own and use certified recovery equipment.

## Q: Is there a penalty if I am caught intentionally venting?

Yes! The penalty can include fines as high as $25,000.00 *per day* per violation and jail terms of up to 5 years. There are several contractors and equipment owners already facing possible fines. As of August 1993 (only a little more than a year after the regulation went into effect), violators had agreed to pay a total of $39,000 in fines, and another $456,000 had been proposed.

| Type of appliance with which recovery or recycling machine is intended to be used | (Manufactured before November 15, 1993) Inches of Hg vacuum | (Manufactured or imported after November 15, 1993) Inches of Hg vacuum |
|---|---|---|
| HCFC-22 appliances, or isolated component of such appliances, normally containing less than 200 pounds of refrigerant | 0 | 0 |
| HCFC-22 appliances, or isolated component of such appliances, normally containing 200 pounds or more of refrigerant | 4 | 10 |
| Very high-pressure appliances | 0 | 0 |
| Other high-pressure appliances, or isolated component of such appliances, normally containing less than 200 pounds of refrigerant | 4 | 10 |
| Other high-pressure appliances, or isolated component of such appliances, normally containing 200 or more pounds of refrigerant | 4 | 15 |
| Low-pressure appliances | *25 | *25 |

*mmHg absolute

*FIGURE 1. Levels which must be achieved by recovery or recycling equipment.*

Probably the heftiest single fine proposed so far came in May 1994. One car shop owner is facing the possibility of a $250,000 fine and 5 days in jail for venting refrigerants during servicing. Another company had a fine of over $105,000 proposed for 10 separate counts of not using recycle/recovery equipment when servicing air-conditioners.

### Q: What is the difference between recovery, recycling, and reclamation?

The regulations relating to refrigerant recycling use the terms *recover*, *recycle*, and *reclaim* to identify different levels of refrigerant processing.

**Recovery** is the removal of the refrigerant from the appliance by use of equipment that captures and contains the refrigerant. This refrigerant is not neces- sarily tested or processed to ensure that it is free of contaminants (such as oils).

**Recycling** is the process of extracting the refrigerant from the appliance and cleaning it for reuse. The contaminants in the refrigerant are reduced by oil separation, noncondensible removal, or single- or multi-stage passes through devices that reduce moisture, acidity and particulate matter. Recycling of refrigerant is usually performed at the jobsite or in a local service shop.

**Reclamation** is the reprocessing of the refrigerant to meet the purity requirements of the ARI-700 standard ("Specifications for Fluorocarbon Refrigerants"), and the verification that this level of purity has been met. Verification can only be done through analytical testing at an EPA-certified testing laboratory.

Generally, you will not need to purchase separate recovery machines and recycling machines. The recycling machines on the market today are capable of recovering the refrigerant directly from the equipment and then recycling it.

## Q: Does my refrigerant need to be "reclaimed?"

In general, the answer is no. Most recovery-recycling equipment currently on the market can clean refrigerants to levels sufficient for reuse in your existing equipment. The regulations allow refrigerant to be recovered, recycled, and reused in equipment belonging to the same owner, without requiring that the refrigerant undergo testing to ensure that it meets the ARI-700 standard. This applies to an equipment owner with several pieces of equipment at the same site, as well as an owner of a chain of stores or other facilities at different locations. However, as of August 6, 1994, recovered refrigerant cannot be sold from one owner to another unless it has been reclaimed—that is, unless it has been certified that it meets the ARI-700 standard. Many refrigerant manufacturers have developed programs for service contractors to return used (nonrecycled or reclaimed) refrigerant through their wholesale-distribution networks in return for credits.[1]

## Q: Do I have to purchase recovery-recycling equipment?

If you are servicing or maintaining refrigeration equipment, or you are a service contractor who provides this service to others, *you must use recovery/recycling equipment when performing service, maintenance, or disposal activities.* If you already own a recovery/recycling unit that you purchased prior to May 7, 1994, and that unit is capable of meeting the vacuum levels in Figure 1, then your equipment can be "grandfathered" into the regulations.* In addition, all homemade recovery/recycling units that were built before August 6, 1993, can be

legally used until they are no longer effective. After that time, any new unit must be certified.

## Q: What are the requirements for recovery/recycling equipment?

Recovery/recycling equipment that has not been grandfathered or was not homemade prior to August 6, 1993, must be certified by an EPA-approved testing organization. In order to be certified, the equipment must be capable of achieving the required evacuation levels. The exception to this is recovery/recycling equipment used on small appliances containing less than 5 pounds of refrigerant (e.g., window air conditioners). These units must recover 80% to 90% of the refrigerant in the system.

## Q: Are there special requirements for system-dependent recovery units?

Yes. System-dependent units (which use the compressor of the equipment being serviced to pump out the refrigerant) are approved for use under the regulations. However, these systems are restricted to use on systems which contain 15 pounds of refrigerant or less. In addition, a backup self-contained recovery unit is required for use on systems with non-operational compressors.

## Q: Are there any requirements to fix leaking systems?

Yes. Industrial and commercial process refrigeration systems containing 50 pounds or more of charge must be repaired if the annual leak rate is 35% or more, unless the owner has developed a plan to have the system retrofitted or replaced within 1 year. All other chillers, including those for comfort cooling, with charges of 50 pounds or more must be repaired if the annual leakage rate is 15% or more, unless the equipment is to be retrofitted or replaced within 1 year. A copy of the plan must be kept at the site of the equipment and be made available for EPA inspection on request. In addition, the equipment owner is required to maintain records of refrigerant purchased and added to the equipment each month.

The requirements for industrial process refrigeration are undergoing revision as a result

---

*"Grandfathering" is the term EPA uses for equipment that was purchased before the regulations went into effect, but which meets the current standards for recovery. To find out what the procedures are to grandfather your equipment, contact your regional EPA office.

of a settlement agreement between EPA and the Chemical Manufacturers Association. This agreement allows a 120-day repair period in cases in which the repairs would require the shutdown of an industrial process. The settlement agreement also allows repair periods of more than a year if delays are due to federal, state, or local regulations—or if suitable replacement refrigerants are not available*. The final requirements for industrial process refrigeration will be published in November 1994.

## Q: Do all service technicians need to be certified?

Yes. Anyone who maintains, services, repairs, or disposes of appliances must be certified by an EPA-approved technician-certification program. Without certification, technicians will not only be barred from servicing, maintaining, or repairing equipment, but they will also be unable to purchase refrigerants from wholesalers or retailers. The certification is only required once and does not have to be renewed. Although there are many training programs available, the EPA does not require additional training as part of certification. Training may be helpful though, especially for technicians who have been out of school for a long time.

## Q: What type of certification is required?

There are four types of certification.

**Type I** is for technicians who maintain, service, or repair small appliances (with 5 pounds or less of refrigerant), such as window air conditioners, home refrigerators or freezers, vending machines, and drinking-water coolers.

**Type II** is for technicians who maintain, service, repair, or dispose of high- or very high-pressure equipment. High-pressure equipment is defined as an

appliance that uses a refrigerant with a boiling point between -50C and 10C at atmospheric pressure. This includes equipment using R-12, R-22, R-114, R-500, or R-502.

**Type III** is for technicians who maintain, service, repair, or dispose of low-pressure appliances. Low-pressure equipment is defined as an appliance that uses a refrigerant with a boiling point above 10C at atmospheric pressure. This includes equipment utilizing R-11, R-113, and R-123.

**Type IV** is a universal certification for technicians maintaining, servicing, repairing, or disposing of both high- and low-pressure equipment.

The regulations prohibit technicians from working on equipment for which they are not certified. Keep in mind, though, that you only need to be certified for the type of equipment that you actually service. It is not necessary to get Type IV certification if you only work on equipment with charges of 5 pounds or less.

## Q: Can I still buy CFC and HCFC refrigerants?

Until the production phaseout for these chemicals goes into effect, new supplies of CFCs and HCFCs should be available. In order to purchase these refrigerants, you need to be an EPA-certified technician. After the December 1995 production phaseout for CFCs, supplies of mostly recycled refrigerants will still be available from manufacturers and retailers as long as the supplies last. Some estimates of CFC shortages have been alarmingly high. In 1993, DuPont estimated that CFC-12 shortfalls could be as high as 81–140 million pounds. Predictions for CFC-11 shortages have reached 1–3 million pounds. Obviously, at some point being able to legally purchase CFCs will no longer be an issue—there will be none left to buy.

## Q: Can I stockpile refrigerants?

Yes, if you can afford to. There are no regulations against stockpiling refrigerants, and both EPA and other organizations have recommended storing refrigerants as a means to ensure that sufficient supplies are available to meet current

---

*It is important to note that EPA's definition of industrial process refrigeration differs from that used elsewhere in this book. EPA defines industrial process refrigeration to include "ice machine and ice rinks, as well as many complex, customized systems used in the chemical, pharmaceutical, petrochemical, and manufacturing industries."

and future needs. The question of stockpiling then becomes more of a financial one than a legal one. In 1991, Congress placed taxes on the sale, use, or importation of CFCs. These taxes have increased the costs of refrigerants by 3% to 10% per year. The tax on CFCs is $4.35 per pound in 1994 and will rise to $5.35 in 1995. The purchase price of CFCs tripled in 1990, and had reached 10 times the pre-regulation levels by 1993. Before you begin stockpiling refrigerants, you need to identify any state or local regulations that may limit the amount of refrigerant you may store on site, or create other regulatory barriers to the development of a refrigerant stockpile.

## Q: Won't my supplier be able to find an alternative for me?

Not necessarily. Representatives from the refrigerant industry have warned users not to expect a "silver bullet" solution or alternative. Most alternatives require at least some retrofit, while others require retrofits so extensive you practically have a new system by the time you're done. There may also be a problem with lead time. If you wait until the last minute, expecting that "silver bullet," you may find that there is a significant lead time before your order can go through. If too many users hold off on retrofitting, the stock of equipment or refrigerant may not be readily available when you decide you need them. Preparation is the key to surviving the phaseout of CFCs. If you wait and assume an easy alternative will be available, you could find yourself unable to continue operating.

Once you have assessed your options, you should contact your supplier for information on refrigerant cost and availability. Your supplier can also help identify alternatives that may be workable for your applications.

## Q: How difficult will it be to replace my refrigeration equipment?

The answer to this question is up to you and how well you plan. Just like waiting to look at alternatives, replacement equipment may not be immediately available when you decide to purchase it. Chiller manufacturers estimate that by January 1996, once CFC production has ended, there will still be about 35,000 CFC-using units left to be replaced or retrofitted. Hence, you will not be alone in needing new equipment and may not be able to get it immediately.

## Q: How do I dispose of my old equipment when I phase out?

If the equipment is dismantled on site prior to disposal, the refrigerant must be removed and recovered. This generally applies to commercial refrigeration units, chillers, and industrial chillers. The ultimate responsibility for ensuring that this is done rests with the equipment owner. The person removing the refrigerant must be certified, and must reduce the system pressure to below 102 millimeters of mercury vacuum using approved, certified equipment.

For equipment that is usually disposed of with the charge intact, such as refrigerators, window air conditioners, etc., the responsibility for removing the refrigerant rests with the final person in the disposal chain. This is usually a municipality, a metal recycler, or a landfill owner. Hence, homeowners are not responsible for removing the refrigerant from an old window air conditioner or refrigerator. Whoever recovers refrigerant from these small appliances must either recover 90% of the refrigerant in the appliance when the compressor is operating or 80% of the refrigerant when the compressor is not operating, or evacuate the appliance to four inches of mercury vacuum.

## Q: What options do I have when selecting replacement refrigerants?

Under the Significant New Alternatives Policy (SNAP) program, the EPA has identified alternative compounds and technologies that are acceptable for replacing refrigerants in most major use categories. The SNAP regulations prohibit users from replacing a CFC or HCFC refrigerant with another chemical that poses a greater risk to human health or the environment than the CFC or HCFC that it replaces. Some of the alternatives that EPA has designated as acceptable for replacing CFCs require use controls to minimize their environmental or human health impacts. A discussion of the available replacement refrigerants and their uses are presented in Chapters 9 and 10 of this book.

## CONCLUSIONS

This chapter has presented some of the most common questions regarding the production phaseout of halocarbon refrigerants, and the ways in which the regulations being implemented under Title VI of the Clean Air Act can affect equipment and refrigerant users. In order to protect yourself and your operations from being adversely impacted by the refrigerant phaseout, you need to ensure that you comply with the regulations for recovering and recycling refrigerant, and you need to develop a plan for transitioning into alternative refrigerants or technologies at some point in the future. The next section will provide information on what you need to know about your existing equipment to make informed choices on future changes.

## REFERENCE

1. *The Air Conditioning, Heating, and Refrigeration News.* Business News Publishing Co: Troy, MI. "A contractor's guide: Complying with EPA's refrigerant recycling rule." Vol. 189, No. 1. May 3, 1993: 15.

# Section II

# KNOW YOUR EQUIPMENT

# CHAPTER 4

# WHAT HAVE WE HERE?

It is clear from the preceding chapter that the production phaseout will lead to some potentially significant impacts for users of CFCs and HCFCs. EPA and the chemical and equipment manufacturers have been expecting to see increased levels of planning and phaseout activity among CFC users since the announcement of the acceleration of the phaseout schedule in December 1993. Remarkably though, users have been slow in developing and implementing phaseout plans.

Many users may be assuming that the regulations of the Clean Air Act simply do not apply to them. After all, the regulations are aimed at eliminating the "production" of CFCs and HCFCs. Some CFC users may be under the impression that they do not need to develop plans for switching to alternative refrigerants now. They may be thinking that there will be no problems finding new refrigerants or obtaining new equipment after halocarbon production ceases. Others may simply be overwhelmed by all the issues they must take into account in identifying alternatives to their CFCs.

The previous chapter should have dispelled any belief that the Clean Air Act has no relevance for CFC and HCFC refrigerant users. EPA has demonstrated that it is serious about prosecuting violators of the venting regulations, and there is little doubt EPA will vigorously enforce the leak prevention requirements as well. In addition to the regulations, there is also an economic incentive for developing a refrigerant-management plan. The predicted shortages of some CFC supplies after the production phase-out date will likely mean that users who wait until after 1995 to develop plans may find themselves unable to obtain the CFCs that they need, thus being forced into a quick, and perhaps costly, decision on replacing their equipment. Even more disconcerting are the predictions that if the majority of users wait until they can't get CFCs to make a decision on retrofitting or replacing their equipment, they may find that they are on a long waiting list for equipment and technical expertise to make these changes.

For all of these reasons, the development of a long-term strategy for refrigerant and equipment management makes sound economic and business sense. The majority of equipment manufacturers, and some of the larger chemical companies, have been developing programs to provide their customers with assistance in developing refrigerant-management plans. There is no single approach to planning that will positively identify the single best alternative for all circumstances. Much of the decision making eventually comes down to site-specific and case-specific requirements. The single best tool that a decision maker should have is a firm grasp of the issues and the alternatives available. And the structure of any plan begins with information about your existing equipment.

## INTUITION

Throughout this section you are going to be asked to review information on your existing equipment. The person who has been regularly working with this equipment is the person best able to answer most of the questions about its current state of health. How do you determine if the equipment is in fair or poor condition? The following chapters will give some guidelines for making this determination, but the final decision is going to lie with the person who knows the equipment best. A piece of equipment may be old but in excellent working condition. On the other hand, the equipment may have been poorly maintained or allowed to run down. An external evaluation of the equipment should be able to give you some "gut feelings" about the internal workings of that equipment.

It is important to remember that the goal of a refrigerant-management plan is not to obtain completely accurate information on all aspects of your existing equipment. The collection of this "perfect" data would be time-consuming and costly. What you are looking for are some benchmarks and estimates for making informed decisions on when equipment should be retrofitted or replaced. From the data collected, owners and operators will need to use their best engineering judgement to evaluate their equipment and determine its health and potential longevity.

# CHAPTER 5

# TAKING STOCK
## (Inventory)

The first step in any plan is to collect information about your current equipment. The choices that you have in identifying potential retrofits or replacements may be limited by the environment associated with your existing equipment. For example, if your system is located in a sub-basement, replacing that equipment may have severe implications for building operation. You may find it necessary to remove walls or other permanent parts of the structure just to get the old piece of equipment out and the new piece of equipment in.

Information on the existing equipment is also necessary for determining the optimal time to retrofit or replace, and which of these two options makes the most sense. Because CFCs have not really been regulated in the past, and because they were relatively inexpensive to buy and use, many equipment owners may not have a very good idea of the health of their existing system. Leakage rates may not have been tracked in the past because they were really not considered a problem. Now, the Clean Air Act is requiring that you know how much your system leaks and that you make plans to have it repaired or replaced if it exceeds the allowable leakage rate. In addition, if your equipment is leaking, even if it is below the regulatory limits, it may be leaking money that you could be saving.

The first step in the "Know Your Equipment" part of planning is the equipment inventory. This inventory will form the basis for general refrigerant management and decisions on whether the equipment can be kept for a while longer, should be retrofitted, or should be replaced. Table 1 is an example of the type of information that should be included in the initial inventory. The information in this table can be used to make a determination of the equipment's remaining useful life, and the current energy requirements of the equipment.

## GENERAL INFORMATION

This portion of the your equipment inventory sheet includes some general information about the equipment, including equipment age, location, manufacturer, model, serial number, and warranty. If you do not own the equipment, you will want to note who does. This may be important when determining who owns and is responsible for properly disposing of the equipment. A sketch or diagram of your mechanical room with individual pieces of equipment identified can be helpful in classifying the location, accessibility, and ventilation status of your mechanical room. This sketch should include the location and types of ventilation, self-contained breathing apparatus, and other safety equipment.

Tied to the equipment's location is the question of how accessible the equipment is. For example, it is important to know if replacing equipment will also involve removing walls, which would substantially increase the initial cost. If the equipment is located in a hard-to-access area, the information recorded on the inventory sheet should include a note about the potential to relocate the equipment to a more accessible area.

This section of the inventory form also has a place for you to record your equipment's condition. A later chapter will discuss the assessment of condition in the framework of determining the equipment's remaining useful life. However,

a general determination of whether your equipment is in good, fair, or poor shape will help you decide whether, and for how long, it is worth keeping around.

## REFRIGERANT INFORMATION

The regulations of the Clean Air Act will require you to keep records on refrigerant usage for all refrigerants, not just CFCs or HCFCs. For this reason, it may make sense for you to inventory all your refrigeration and air conditioning equipment, including those pieces that use ammonia or other non-halocarbon refrigerants. You will want to record the type of refrigerant that the system uses as well as the amount of charge. In addition, for the refrigerant CFCs and HCFCs that you use, you will also want to record the amount of refrigerant that you currently have in storage. Tracking how much of each refrigerant is in each piece of equipment and in storage will help you determine how much refrigerant in the various systems is potentially available for recovery, recycling, and reuse.

## EFFICIENCY INFORMATION

This information is provided to give you an idea of the energy use and efficiency of your equipment. In a later section, we will discuss how energy efficiency considerations will be important in determining whether to retain, retrofit, or replace your equipment. For HVAC equipment, information on equipment cooling capacity will also be important in determining whether you are meeting or exceeding the cooling load requirements of your building.

Like many equipment owners, especially if you use industrial process refrigeration, you may not know the actual energy consumption of your equipment. Unless the energy flow to the equipment is metered, there will be no direct method for determining how much energy your equipment requires to run. A good estimate of energy consumption can be calculated from the coefficient of performance of the equipment and the annual operating hours. The coefficient of performance (COP) can be used to calculate the kilowatts per ton (kW/t) of energy used, and is generally provided by the manufacturer for every new piece of equipment. If you can get the kW/t figure from your equipment information, you should use this number on your equipment inventory form as it will save you from having to convert from COP. You may have to dig up the information on your equipment to find this information.

## LEAK INFORMATION

The sample inventory sheet (see Figure 1) contains a section for information regarding leaks for each piece of equipment. Criteria that will help you evaluate the leak rate for equipment include: pounds of refrigerant added annually to the equipment, the extent of the piping, the type of purge units, general condition of the equipment, and its maintenance history. A discussion of the importance of leak detection and maintenance is provided in the next chapter.

## EQUIPMENT INVENTORY

Technician's Name: _____

Telephone Number: _____

Date: _____

---

**GENERAL INFORMATION:**

Building _____

Location/Room _____

Type of Equipment _____

Equipment Model _____

Serial No. _____

Manufacturer _____

Age of Equipment _____

Warranty?    Yes    No    Through: _____

Easily Accessible?    Yes    No

   Explain _____

General Condition of equipment _____

**REFRIGERANT INFORMATION:**    Type of Refrigerant _____

Charge _____ Lbs

**EFFICIENCY INFORMATION:**

Equipment Capacity _____ Tons

Estimated Motor Efficiency _____

Estimated Energy Consumption _____ kW/hr

Estimated Annual Hours of
   Operation at Full Load _____ Hrs/Yr

Coefficient of Performance _____

Kilowatts per ton _____

**LEAK INFORMATION:**

Lbs of Refrigerant Added Yearly _____

Extensive Piping?    Yes    No

   Explain _____

Type of Purge Unit _____

Estimated Leak Rate _____ % yearly

History of Leaks?    Yes    No

   Explain _____

*FIGURE 1. Equipment inventory.*

# CHAPTER 6

# LEAKING AWAY THE FUTURE
## (Leak Prevention)

There are several reasons for wanting to know the leak rate, and leakage potential, of your existing equipment. One of the strongest reasons was addressed in Chapter 3, under the discussion of the need to repair equipment that exceeds the allowable leakage rate established by EPA in May 1994. The regulations currently require industrial and commercial process refrigeration systems containing 50 pounds or more of charge be repaired if the annual leak rate is 35% per year or more. All other chillers, including comfort cooling, with charges of 50 pounds or more must be repaired if the annual leakage rate is 15% or more. Optionally, equipment owners may decide to have the equipment retrofitted or replaced. A plan for retrofit or replacement must be developed within 30 days of discovering that the equipment exceeds the maximum leak rate, and the retrofit or replacement must be completed within one year.

Even if your equipment does not exceed the maximum allowable leak rates, it is important to know how much refrigerant you are losing to leaks in order to determine what your future requirement for refrigerant will be. A primary goal of refrigerant management planning is to identify what type and quantities of refrigerant will be necessary to service your existing equipment, should you decide not to retrofit or replace it immediately, and to determine whether those quantities will actually be available. Knowing how much it currently costs you to replace refrigerant lost by leaks can also provide incentive for leak repair and management investments. Refrigeration and air conditioning systems are designed as sealed units, and they work best when that seal is intact. Therefore, even if there were no regulations, repairing significant leaks can result in more efficient operation of your equipment, thereby saving energy and money.

## ESTIMATING LEAKAGE

For some equipment owners, estimating the annual leakage rate will be as easy as identifying the amount of refrigerant purchased per year or the number of service calls and the amount of refrigerant added. Those purchases can be divided by the total amount of refrigerant in the system to determine the percent of refrigerant that has leaked.

If refrigerant is purchased in bulk for more than one piece of equipment, or if records of refrigerant purchases are not available, it will be more difficult to determine the actual leakage rate of the equipment. Estimates can be made using the average annual leakage rate for the category of equipment. This rate is estimated to be 9% per year for industrial refrigeration, and 30% per year for commercial refrigeration equipment.[1] The average estimated leakage rate can be used to calculate the amount of refrigerant that will be required to service existing equipment over the remainder of its useful life.

Individual pieces of equipment will have different leak rates, depending on their condition, the extent of piping on the system, types or existence of purges, etc. For example, in many retail food cooling and display applications, extensive refrigerant lines running from the condensing units to evaporator coils in the display cases with numerous fittings and connections tend to create a high leakage potential. In some of these applications, annual leakage rates may be as high as 60%.[2] Moreover, simply knowing the mount of leakage will not tell you where those leaks are occurring. In order to determine the actual amount that the system is leaking and where leak repair is required, the system will need to be inspected.

## LEAK DETECTION

Equipment is generally manufactured and installed to minimize refrigerant leakage. Over time, however, leaks can develop as a result of normal wear, improper installation, improper operation, and inadequate or improper service and repair. The primary method for identifying leaks is the routine inspection using any number of leak-detection devices. Leaks are most often found in association with tubing, flanges, o-rings, and connections where components meet. Leaks can also occur through loose fittings, welded joints, and worn gaskets and seals. Visual inspection can often identify signs of potential leaks. For example, leaking lubricant can indicate a possible refrigerant leak as well. An undercharge of the system might also indicate a problem.

The American Society of Heating, Refrigerating, and Air-Conditioning Engineers (ASHRAE) has developed leak-detection guidelines for systems using halogenated refrigerants. ASHRAE Guideline 3-1990 and Guideline 3a-1992 provide an overview of the types of detection equipment available for use on halogenated refrigerants. These guidelines also provide general approaches for leak detection.

One advantage of the Clean Air Act's requirements to contain refrigerant leaks, and the increased regulation of CFC and HCFC refrigerants, is the added incentive for manufacturers of leak-detection equipment to develop more sensitive techniques for pinpointing leaks. The main approaches to leak detection include visual inspections, use of portable detectors, or automated tests.

## LEAK REDUCTION

There are several options for reducing future leakage at points in the system where these are most likely to occur. Replacing worn gaskets and seals, and tightening loose fittings are some obvious activities that can reduce system leaks. The installation of leak-detection equipment, or upgrading that equipment to be more sensitive, can also help in leak control. These systems can help identify the need for a leak inspection while ensuring that refrigerant leakage levels do not exceed safety allowances.

The first step in identifying areas to tackle for leak reduction is to put together a list of all the potential leakage points on your equipment. Add to the list maintenance or service practices that may also provide opportunities for refrigerants to leak. Some of the information that you may wish to include in this list is provided in Figure 1. Once you have made a list, you can inspect your equipment on a regular basis and schedule leak repair activities as the need arises. An example of a schedule suggested by ASHRAE is found in the textbox.

Low-pressure equipment can be retrofitted with high-efficiency purge units. Because air leaks into the low-pressure side of chiller equipment, rather than refrigerant leaking out, purging is one of the most common causes of refrigerant leakage. Low-efficiency purges lose refrigerant at the rate of approximately 4 lbs of refrigerant for every 1 lb of air. In some of the newer purge systems the ratio is now down to 1 lb of refrigerant for 1 lb of air. Some new purges are now being advertised that lose only 3/4 oz of refrigerant to each pound of air purged[3].

Low-pressure chillers used for comfort cooling may also have leakage problems if they are only used on a seasonal basis and not properly prepared for the off-season. Air and moisture can leak into the system when it is not running, necessitating a large purge of the system on start-up. To prevent this, nitrogen can be added to the system during periods of non-use to bring the pressure above ambient or outside levels. This will reduce the amount of air and moisture leaking in, while minimizing the degree of system purging required.

## RECORDKEEPING

The regulations governing leakage rates require equipment owners to keep records of equipment-servicing activities. This information will need to include the date of the service call and the amount of refrigerant added to each piece of equipment. For equipment owners who service their own equipment, the records will need to include the amount of refrigerant added to the equipment each month.

**Certain procedures can be monitored to eliminate most leaks:**
- Evacuation
- Charging refrigerant
- Transfer of refrigerant to or from recovery device
- Transfer to or from receivers
- Seal problems
- Purger problems
- Refrigerant handling and storage
- Lack of proper leak testing

**Other specific sources of leaks which can be eliminated include:**
- Screwed or brazed piping
- Instrument or tubing leaks
- Receiver leaks
- Rupture disks/relief valves
- Valve packing
- Expansion joints
- Gaskets
- Valves on cylinders
- Rusty piping
- Sight glasses
- Misoperation
- Balance in rotors
- Intentional blowdown (to remove air)
- Charging hoses
- Improper storage
- Refrigerant left in large cylinders
- Missing valve caps
- Inaccurate measuring devices
- Remote instruments (e.g. panel board)
- Deteriorated "O" rings
- Defective parts
- Rounded valve stems
- Line breakage
- Improper evacuation prior to maintenance
- Overcharging
- Wrong equipment in acid areas
- Scattered equipment and not enough time allowed by production to make proper repairs
- Multiple people/crews involved in refrigerant transfer
- Refusal to shut down a leaking piece of equipment of equipment for repairs
- Lack of training
- Changing refrigerant filters
- Deterioration of shell
- Flanges
- Air leaks

FIGURE 1. List of potential leakage points.
(Source: Engineered Systems, Nov/Dec 1992 Tech-R-108— Take Action Today—Tenessee Eastman's suggestions)

The goal of this recordkeeping requirement is to provide information to determine leakage rates and to ensure that these rates do not exceed the maximum allowable rate established by the regulations. Most equipment service contractors already maintain records of this type, as do most providers of recovery and recycling services. Equipment owners may have a system for tracking this information to ensure that equipment is in compliance with the regulations. Figure 2 is an example of the type of information that can be collected to track refrigerant-maintenance activities.

## CONCLUSIONS

Refrigerant containment is one of the most important aspects of overall refrigerant management. In addition to regulatory requirements to keep leakage below certain established levels, there are also economic reasons for wanting to minimize the amount of refrigerant being lost that will need to be replaced. The discussions presented in this section will have applicability for your replacement refrigerants as well. Concerns over the ozone-depletion potential and global warming potential of refrigerants are directly related to the ability of the refrigerant to be emitted. The tighter your system, the more efficiently it will operate, the more money you will save by avoiding refrigerant replacement costs, and the more environmentally friendly your operations will be.

## REFERENCES

1. United Nations Environment Programme, Montreal Protocol 1994/5 Assessment, Refrigeration, Air Conditioning and Heat Pumps Technical Options Committee, Draft Chapter 7, *Air Conditioning and Heat Pumps.* July 1994. Nairobi, Kenya.

2. Salas, C.E. and M. Salas. 1992. *Guide to Refrigeration CFCs*. Fairmont Press, Inc., Lilburn, GA. pg. 238.

3, Trane Company, Technical Information Brochure, High Efficiency Purge Systems. June, 1994. *Trane CFC Toolbox,* Trane Company, Arlington, VA.

Technician's Name: _____
Telephone Number: _____
Date: _____

| | | | |
|---|---|---|---|
| *GENERAL INFORMATION:* | Building | _____ | |
| | Location/Room | _____ | |
| | Type of Equipment | _____ | |
| | Equipment Model | _____ | |
| | Serial No. | _____ | |
| | Unit Identification | _____ | |
| | Manufacturer | _____ | |
| | Refrigerant Type | _____ | |
| | Design Refrigerant Charge | _____ | |
| | | | |
| *LEAK IDENTIFICATION:* | Leak Detector Used | _____ | |
| | Leak Location | _____ | |
| | Leak Repaired | Yes        No | |
| | If No, why not? | _____ | |
| | Amount of Refrigerant Added | _____ | |
| | General Comments | _____ | |
| | | _____ | |
| | | | |
| *RECOVERY/RECYCLE* | Equipment Used | _____ | |
| | Recycled? | Yes        No | |
| | Disposition | _____ | |
| | Amount Recovered | _____ | Lbs |
| | Re-installed in Same System? | Yes        No | Lbs |
| | Amount of Refrigerant Added | _____ | Lbs |
| | General Comments | _____ | |
| | | _____ | |
| | | | |
| *UNINTENTIONAL VENTING:* | Venting Situation, Explain | _____ | |
| | | _____ | |
| | Approximate Amount Vented | _____ | Lbs |
| | Amount of Refrigerant Added | _____ | Lbs |
| | General Comments | _____ | |
| | | _____ | |

| | | | |
|---|---|---|---|
| *SUMMARY:* | Refrigerant Used | _____ | |
| | Total Amount Added | _____ | |
| | Total Amount, as % of design refrigerant charge | _____ | |
| | Comments | _____ | |
| | | _____ | |

*FIGURE 2. Refrigerant log.*

# CHAPTER 7

# THE CLOCK IS TICKING
## (Remaining Useful Life)

Another important component of knowing your existing equipment is determining what condition it is in and what the expected remaining useful life of the equipment is. Many approaches to refrigerant management use the existing equipment's age as a determining factor in whether the equipment should be retrofitted or replaced. Equipment age is an important component, of course, but it does not tell the whole story of how healthy the existing equipment is and whether replacement would actually be cost-effective.

What is meant by "remaining useful life"? Under many management plans, it is the estimated amount of time that the equipment can continue to be operated. This is based on an average lifetime of 25-30 years for commercial and industrial process equipment. Many equipment owners have systems that are significantly older than this, and some of that equipment may be in relatively good condition. In general, however, equipment manufacturers recommend that you make plans to retrofit or replace your equipment if it is over 15 years old.

In addition to the potential lifetime of a piece of equipment, it is also necessary to have some indication of the overall condition of the equipment. While it is generally true that a newer piece of equipment will be in better operating condition than an older one, a relatively new piece of equipment may not in fact be cost-effective to maintain if it was improperly installed and leaks excessively, if it is in poor condition because of neglect or environmental circumstances, or if it is not maintaining the desired level of efficiency. Some of these problems may be resolved by proper maintenance and repair. However, an important question is whether it is cost effective to repair the equipment, given the requirement that it will eventually have to be replaced.

The best guideline for determining the condition of your system is your knowledge of its maintenance and service history. Also important is the leak history and leakage data that you reviewed in the previous chapter. A system that requires a significant amount of service and has a history of severe leakage should probably be listed in poor condition for the purposes of determining the remaining useful life. Another consideration in determining the remaining useful life of the equipment is the number of changes that have been made to it over time. The nature of these systems is that they are made up of component parts. It is possible that the system you currently have has had some of these parts replaced fairly recently. In determining the equipment's age, then, it is important to keep in mind that some of the system components may actually be younger in age than the overall system.

You can liken this situation to that of owning an old car—like a 1970 VW Bug. By the time we reach 1994, it is more than likely that you have replaced the engine, perhaps the transmission, maybe even made some repairs to the body. In 1994, you have a car that was originally built in 1970, but the main components of that car, the ones that keep it running, are all more current than that. The same may be true of your refrigeration or air conditioning equipment. When assessing the condition, take a look at both the inside and the outside of your equipment.

The potential lifetime of the equipment will also be affected by the amount of time that the equipment is in operation. Equipment that is operated at full load, 12 hours a day, 365 days a year will require more maintenance and generally have a shorter lifetime than equipment that is operated seasonally or only during working hours. As with assessing the condition, it is necessary to consider the costs to maintain this equipment, or to make major repairs, in the face of eventual replacement.

The goal of determining the equipment's remaining useful life is two-fold: it provides a time frame in which decisions on retrofit and replacement must be made, and it provides a time frame for estimating the amount of refrigerant required to maintain the system in the future. It is important that the determination of remaining useful life accurately reflect the amount of time the equipment can continue to be used effectively, without unnecessarily prolonging the life the equipment.

A general approach to determining the remaining useful life is to subtract the equipment age from the estimated lifetime of the equipment (generally 25 to 30 years). You can then adjust this number to reflect the equipment's condition and the number of hours that it is operation. For example, if the equipment is in poor condition, you are likely not getting the performance out of it that you need or would like. Over time, the condition is also likely to go from bad to worse. Try to make a realistic assessment of how much longer you can really keep that piece of equipment operating without expending a significant amount of time in maintenance and repairs. You may find that you want to subtract as much as 5 years from the amount of time you could theoretically keep the equipment operating based on the estimated equipment lifetime.

There are several other components of the remaining useful life that can be considered. For example, the current efficiency of the equipment can be important in determining whether the equipment is meeting current requirements. Efficiency is very difficult to calculate for existing equipment because of the need for very accurate temperature measuring instruments and an accurate method to measure flow rates—which are difficult to measure precisely in the field. Under most situations, the manufacturer's "name plate" which provides information about system components (pump motors, etc.)

will need to be used to provide the efficiency information.

## CONCLUSIONS

In the final analysis, the estimate of the remaining useful life will require a sound judgement based on all available information. The estimate will be primarily based on the best equipment history available, the personal knowledge of the personnel maintaining the equipment, and the art of "Best Engineering Judgement."

The remaining useful life of building comfort cooling equipment may also be influenced by the question of how well it is meeting your needs. Equipment that is significantly oversized or undersized can have reduced energy efficiency or not meet the comfort requirements of your tenants or workers. In the next chapter the issues involved in cooling load calculation are discussed. Meeting the cooling load requirements is something that must be considered in the proper management of the equipment systems.

# CHAPTER 8

# LOSING YOUR COOL?
## (Cooling Load)

While you are getting to know your equipment, take a close look at your cooling load requirements. Estimating the cooling load will help you decide if you have the right type and capacity equipment for your building or space. Over time, your cooling needs may have changed. For instance, if there is more physical activity or more people are using a particular space than in the past, your cooling load requirements will probably have increased. Conversely, if the lighting has been extensively updated to a more energy-efficient system, your cooling load, in turn, will probably have decreased. If your cooling needs have changed, you may want to consider either increasing or decreasing the cooling capacity of your equipment. Equipment that is over-sized will generally run at only partial load which may be inefficient. Equipment that is well over-sized will shut down the cooling before the latent heat (moisture) can be removed. This will leave the space cool but very humid. This is one instance where "overcompensating" could hurt, rather than help, you. If the equipment is undersized, the building may not be cooled sufficiently to create a comfortable working environment, and could lead to more frequent breakdown of office equipment (computers, copiers, etc.) due to over-heating.

It is important to remember that the cooling load for a building is dependent on many different criteria. Office buildings, schools, residential homes, hotels, and theaters all have very different operating environments and hence have vastly different cooling requirements. These building uses all house different equipment and numbers of people who perform at various activity levels. They also require different environmental conditions (e.g., temperature and humidity) and have different structural form and materials. This chapter provides guidelines and approaches for commercial office buildings.

When the building was designed, the engineers and architects rigorously calculated the anticipated cooling load requirements based on the anticipated use of the building space. To accurately calculate cooling load requirements on an existing building is a complex undertaking. The goal of this chapter is not to show you how to perform these calculations, but to provide examples of building changes that may have affected your cooling load requirements. If, after reading this chapter, you feel that your cooling load requirements may have changed significantly, you may want to arrange to have accurate cooling load measurements calculated. This may be accomplished through contracting for the services or using in-house resources familiar with doing these types of calculations.

## ESTIMATING YOUR COOLING LOAD

As a very basic guideline, for commercial office buildings, your equipment capacity should generally be one ton of cooling per 350 to 600 square feet. Heat can enter a room 1) through walls, floors, and ceilings; 2) through windows; 3) from occupants; and 4) from infiltration and ventilation. Each of these heat sources are considered when making cooling load calculations. If there have been changes made to any of these source (e.g., the installation of thermal windows), you can expect that the cooling load requirements have been affected. In order to determine whether your

cooling load calculations should be closer to the 350- or the 600-square foot end of the scale, we have provided a review of the major heat sources and their impacts on cooling load. In general, decreases in cooling load will move your calculations closer to the 600 square foot end of the scale, while increases will move you back toward the 350 square foot end.

When considering your cooling load, keep in mind there are two types of heat that affect the cooling load. *Sensible heat* is directly added to the air from its source. In other words, the effects of this heat are immediately felt by the air (like turning on an electric heater). *Latent heat*, as mentioned earlier, is moisture that is added to the air and usually has a delayed effect on the air temperature. For example, in residential homes where people shower, do laundry, cook, etc., the latent heat gain will be more than in a small office environment. All cooling equipment has maximum capacity for removal of both sensible heat and latent heat.

## HEAT SOURCES

### Lighting

Lights, including ceiling lights, desk lamps, lit vending machines, and any specialized lighting, are usually a major source of heat in large offices. Particularly in office buildings, lighting has a significant impact on the cooling load requirements. Usually, the more efficient the light, the less heat it releases into a room. As a reference, very old, inefficient lights can use up to 4 watts/square foot. If a retrofit has been done to replace those lights with newer, more efficient lights, you can reduce that value to 1 watt/square foot. This 3 watts/square foot change could translate into cooling load reductions of as much as 1 ton per 1175 sq ft.[1]

If the lighting in your building has been upgraded to be more energy efficient, you may have experienced a reduction in your cooling load requirements. By estimating the number of lights that have been upgraded, you can get a feel for how large an effect the change has on the internal load. Upgrading many lights can significantly decrease your internal load, and will impact your cooling load requirements. The use of other energy saving devices, such as occupancy sensors, will reduce amount of time that the lights are on and will also decrease the heat that the lights put out.

Lights also release some heat in the form of radiation, which has a delayed effect on room temperature. Rather than being immediately felt as added heat, this radiation is absorbed by the walls, floors, and furniture in the room. Once these structures absorb enough heat and are warmer than the air in the room, they will radiate the heat back out into the room (a process called *convection*). Obviously, this form of heat is difficult to measure. However, you can assume that major upgrades to the lighting will tend to decrease this convection effect.

## Appliances

When estimating cooling load, you should also consider if there have been major changes in the types of appliances and office equipment used in the building. As you would expect, kitchen appliances contribute a great deal of heat in places such as restaurants, schools, hotels, and hospitals. Computers, printers, copy machines, calculators, and sometimes posting machines, check-writers, etc. are the most common sources of heat transfer in offices. The heat generated by electronic equipment in computer areas usually ranges from 75 to 175 Btu/h*ft.[2] For manufacturing plants that perform general assembly, the heat transfer from equipment ranges from 20 Btu/h*ft[2]. Since more and more production is becoming automated today, it is not uncommon to find high concentrations of heat near equipment areas.[2]

Major changes in quantity and type of equipment used in the area will affect the cooling load requirements. These changes may be the result of changes in building occupancy, or they may be the result of increased automation. Whatever the cause, if you suspect that the amount of equipment used in your building has increase, you can assume an increase in your cooling load requirements.

## People

People, also give off heat and moisture during daily activity. In fact, people are a significant source of both sensible and latent heat gain.* For offices where individuals are doing light work, and which are occupied by approximately one person per 100–150 square feet, you can generally assume each person is giving off 250 Btu/hr of sensible heat and 200 Btu/hr of latent heat. A change in building ownership or occupancy may have increased the level of human activity or the number of people occupying the building. Doubling the number of people in offices originally meant for a single person and adding of workspaces in hallways and corners are well-known methods for increasing workspace occupancy. If a significant increase in occupancy has occurred, you can assume that there has been an increase in the cooling load requirements for the building.

Occupancy can also change over time. Areas that were originally anticipated as office space may be converted to storage or meeting rooms that are used on an occasional basis. Changes like these that reduce occupancy per square foot will have a corresponding decrease in the cooling load requirements.

In determining how to factor occupancy into your cooling load estimates, it is important to identify long-term trends versus short-term occupancy shortages or gains. Your cooling equipment will have a significant life-time — generally 25 to 30 years or longer. A decision to increase or decrease cooling load requirements based on occupancy should be based on long-term analysis of future building use.

## Heat from windows

Sunlight shining through windows will also affect the air temperature in a room. Even windows that are always covered by shades allow some heat transfer to occur. The number, size, and type (e.g., double hung, metal sash, etc.) of windows will determine the amount of heat transfer occurring. The addition of energy-efficient thermal windows will decrease the amount of heat entering the building through this route, and will decrease the cooling load requirements. New construction near the building may have also decreased the amount of thermal energy entering through windows by shading some of the windows from the sun. Shading could also indirectly lower the cooling load requirements by reducing the temperature dif-

---

*While humans give off both latent and sensible heat, transfer of latent heat is considered to be instantaneous from humans. This is different from our other discussions of latent heat, which usually has a lag time before it affects the air.

ference between outdoor and indoor air, which reduces the amount of heat that enters the building through conduction.

## Infiltration from outdoors

Heat infiltration from cracks and spaces in windows and doors will also affect the cooling load requirements. Doors that are regularly propped open can allow heat to come in (or cooling to go out!). The tendency on the part of your occupants to leave doors open may indicate that the building is not providing sufficient cooling or air flow. The drawback to this is that your cooling equipment is forced to work even harder to make up for the cooled air that escapes through the door.

Older buildings may have developed spaces around windows that will allow heat to infiltrate the building. You can get a good idea of infiltration from windows by examining the crack around the perimeter of your windows and how much air comes in on a windy day. Here again, construction around your building may have increased the velocity of the wind hitting your windows. This can cause an increase in the amount of outside air infiltrating your building. On the other hand, nearby construction may have also sheltered your building from these winds, causing a decrease in infiltration.

Another way heat is introduced to a building from the outdoors is through ventilation. Local codes and ordinances usually specify certain ventilation requirements for pubic places and industrial installations. For public places and offices, the ventilation requirements usually are a function of the number of occupants and materials in the space. Industrial installations may require more ventilation, depending on occupancy and the industrial processes involved. Ventilation systems that have become clogged or damaged may provide less air transfer and will negatively affect your cooling load requirements.

## Heat transfer through interior walls, floors, and ceilings:

It is very rare that changes to your building will result in a change in the amount of heat transferred through walls, floors, and ceilings. One major way that this can happen is through the construction of additional offices within an already designed space. Sub-dividing the floor space beyond what was called for in the original designs can result in changes in the air flow through the conditioned space and affect your cooling load requirements. An example of this is the situation where some offices are over-cooled while some are under-cooled.

## THE FUTURE

The final criteria you need to consider when evaluating cooling load is how the function of the building will change in the future. Remember, your equipment will have a useful lifetime of anywhere from 25–30 years. If significant changes are planned, such as bringing in a new department, bringing in new capital equipment, major lighting efficiency improvements, window upgrades etc., you should try to factor these changes into decision regarding the cooling load requirements for your retrofitted or replaced equipment.

## REFERENCES

1. Gordon, Harry T. Principal, Burt, Hill, Kosar, Rittelman & Assoc., Consulting Architects and Engineers. Personal Communication. June, 1994.

2. American Society of Heating, Refrigerating, and Air Conditioning Engineers. *ASHRAE 1993 Fundamentals Handbook*. ASHRAE, Atlanta, GA.

# Section III

# CHOICES AND OPTIONS

# CHAPTER 9

# COOL CANDIDATES
## (Alternatives)

It is not necessarily true that equipment users need to understand the properties of various refrigerants to be able to identify a replacement for their CFCs or HCFCs. All refrigerants that are currently available commercially provide satisfactory performance within their specific application ranges. In addition, equipment manufacturers, chemical companies, and even utilities are often quite willing to provide information and advice on how to decide on which refrigerant alternative or technology to select. Nor will an equipment user be required to determine whether the material out of which the equipment is made is compatible with a specific compound, or even which lubricant will work best with which refrigerant.

What an equipment owner *does* need to be concerned with are those issues associated with a specific refrigerant type or technology that may affect the decision to retrofit or replace, or may have implications for workplace safety. For this reason, it is important to know some of the properties associated with refrigerants used in vapor compression and other systems. No single alternative will be suitable for all uses. The equipment owner needs to identify the best option, and balance the tradeoffs, for the particular application.

The primary considerations for equipment users are as follows:

**Ozone-Depletion Potential (ODP):** This is important because it is the ozone-depleting properties of CFCs and, to a lesser extent, HCFCs, that is creating the need to find alternatives to these chemicals. Currently, the primary alternative refrigerants of concern are the HCFCs, which themselves are targeted for production phaseout by 2030. It is important to keep in mind that HCFCs are viable interim replacements for CFCs in many applications, and are currently being used extensively. The timeline for phaseout of these compounds is adequate for users to achieve a return on investment for most HCFC-using equipment. The fact that these substances are being targeted for phaseout, however, should be noted in concert with other considerations regarding CFC replacement. In addition, HCFCs should be included in long-term refrigerant-management planning because eventually HCFC equipment will also have to be retrofitted or replaced.

**Global Warming Potential:** This is also an important issue for any user to consider because of the high level of interest currently being paid to global warming internationally. There are really two issues to consider here: the *direct* global warming potential that results from a compound's ability to trap radiant energy, and the *indirect* impact associated with the energy requirements of equipment that use these compounds. Certain CFC alternatives, including HCFCs and HFCs, as discussed in Chapter 1, are greenhouse gases and function to directly trap radiant energy (or heat), which may contribute to global warming. However, an indirect contribution to global warming can also occur as a result of the energy consumption of equipment using these refrigerants. The energy supplied to this equipment can be derived from fossil fuel sources that emit carbon dioxide ($CO_2$), a highly potent greenhouse gas. The indirect global-warming potential may actually be much greater than the direct global-warming potential, depending on the end-use application being considered as well as the source of energy for

that application (e.g., electricity produced by solar, hydroelectric, or wind energy does produce as much $CO_2$ as coal-burning plants do).

Together, the indirect and direct global warming potential is termed the TEWI, or Total Equivalent Warming Impact. The general rule of thumb is that the more energy-efficient a refrigerant is, the lower the indirect global-warming potential would be. At this time, direct global warming potential is considered by the EPA when it reviews halocarbon refrigerant alternatives; however, it is only one of many factors. Information on direct global warming potential is provided here as a basis for comparing alternative refrigerants, and because international attention is focusing on the global-warming issue with potential implications for users of chemicals with high global-warming potential in the future.

**Material Compatibility:** This issue is especially important for retrofits, as many alternative refrigerants have compatibility concerns with lubricating oils, seals, hoses, and motor insulations, as well as the metals used to fabricate heat exchangers, evaporators, and compressors. The more incompatible a substance is with the existing equipment, the more expensive and labor-intensive the retrofit is likely to be.

**Safety and Health:** This is one of the most important considerations for a refrigerant. Many safety codes limit the use of flammable and toxic refrigerants. The issues of refrigerant safety, however, cannot be separated from the safety aspects of the equipment. Some safety risks, such as flammability, may be countered by equipment design or equipment location. The equipment owner needs to know what toxicity and flammability issues are associated with the refrigerant in order to ensure that the chosen alternative or process meets shop or union safety requirements. ASHRAE has developed the following classifications for refrigerants, based on their toxicity and flammability[1]:

*Class A* are refrigerants for which toxicity is not a factor at concentrations equal to or less than 400 ppm.

*Class B* are refrigerants for which toxicity is evident at concentrations below 400 ppm.

In addition to these toxicity groups, ASHRAE has also designated refrigerants according to flammability, using the following number system:

*Class 1* refrigerants are essentially non-flammable.

*Class 2* refrigerants are moderately flammable.

*Class 3* refrigerants are highly flammable.

Toxicity is generally represented in Threshold Limit Values (TLVs), expressed in parts-per-million volume concentrations in air. These are the maximum conditions, established by the American Conference of Governmental Industrial Hygienists (ACGIH), under which workers can be repeatedly exposed to a substance without suffering adverse health effects. For many of the new refrigerants and refrigerant blends, the TLV has not been calculated*.

Flammability is characterized by the lower flammability limit (LFL) defined by the American Society of Testing and Materials (ASTM) Standard E 681-85. The LFL for refrigerants is the minimum concentration of the refrigerant that is capable of creating a flame when mixed with ambient air at ambient temperatures. As with the TLV, the LFL has not been calculated for some of the newer refrigerants and blends.

The 1994 update of ASHRAE Standard 15 will require the use of refrigerant monitors/sensors in all mechanical rooms regardless of the refrigerant being used. The use of oxygen-depletion sensors are no longer required.

## ALTERNATIVE REFRIGERANTS**

The proposed replacements for CFCs have generally been other hydrogen-containing compounds that are similar to CFCs. This category contains the HCFCs, which are also targeted for phaseout and therefore are considered interim solutions, and the hydrofluorocarbons (HFCs). In addition, some attention is being given to hydrocarbons and natural refrigerants, such as ammonia, in specific applications. This section will discuss the major commercially available refrigerants, the applications in which they are or can be used, and the issues of ODP, GWP, materials compatibility, and safety and health associated with each substance. Table 1 shows the alternative refrigerants and some of their characteristics.

## HYDROCHLOROFLUOROCARBONS (HCFCs)

HCFCs are currently used in many applications, particularly air-cooled vapor-compression air conditioners and heat pumps. Because of their high ozone-depletion potentials, relative to other compounds, the HCFCs themselves have been targeted for phaseout in the United States by 2030. A faster schedule for this phaseout has been discussed in Europe and may be brought before the Montreal Protocol parties for possible adoption. In spite of this, HCFCs are still considered viable replacements for the CFCs currently in use. With some exceptions, HCFCs have a significantly lower ozone-depletion potential than CFCs, and therefore are being recommended as viable interim replacements for CFCs by the EPA. In addition, the time frame in the United States for phasing out HCFCs is sufficient to allow equipment purchased today to meet its useful life.

## R-22

**Process:** R-22 is the most common refrigerant for use in air-cooled vapor compression air-conditioners and heat pumps. Recent estimates from the United Nations Environment Programme indicate a total worldwide inventory of 383,000 metric tons of R-22 equipment.[2] R-22 is also used in most positive-displacement, water-chilled air conditioning applications, and in some centrifugal and screw water chillers. It is being used extensively as an interim replacement for R-11 and R-12 in new chillers.

**ODP:** 0.055

**Direct GWP:** 1600

**Materials Compatibility:** R-22 operates at higher pressure and temperature levels than R-12. The high discharge gas temperature may cause problems involving the thermal

---

* TLV is a trademark of the ACGIH and refers to exposure limits established by that group. Several refrigerant manufacturers have developed indices similar to the TLV for testing their refrigerants.

** Information on refrigerants in this section was obtained from Material Safety Data Sheets, manufacturer marketing information, and other sources.

| Refrigerant | Formula/Symbol | Application* | ODP | GWP | TLV (ppm) | LFL** (%) |
|---|---|---|---|---|---|---|
| CFC-11 | $CCl_3F$ | L | 1.000 | 3400 | 1000 | NONE |
| CFC-12 | $CCl_2F_2$ | HH H M (L) | 1.000 | 7100 | 1000 | NONE |
| CFC-502 | R-22/115 (48.8/51.2) | (H) M L | 0.283 | 5534 | 1000 | NONE |
| CFC-500 | R-12/152a (73.8/26.2) | H | 0.738 | 5254 | 1000 | NONE |
| R-22 | $CHClF_2$ | H M L | 0.055 | 1500 | 1000 | NONE |
| R-134a | $CH_2FCF_3$ | HH H M (L) | 0.000 | 1200 | 1000 | NONE |
| R-123 | $CHCl_2CF_3$ | M L | 0.020 | 85 | 10 | NONE |
| R-401A | R-22/152a/124 (53/13/34) [MP39]† | H M | 0.037 | 1018 | 800 | NONE |
| R-401B | R-22/152a/124 (61/11/28) [MP66] | L | 0.040 | 1116 | 840 | NONE |
| R-402A | R-125/290/22 (60/2/38) [HP80] | (M) L | 0.021 | 2648 | 1000 | NONE |
| R-402B | R-125/290/22 (38/2/60) [HP81] | M L | 0.033 | 2252 | 1000 | NONE |
| R-404A | R-125/143a/134a (44/52/4) [HP62 and FX-70] | n/a | 0.000 | 3520 | 1000 | NONE |
| R-407A | R-32/125/134a (20/40/40) [Klea60] | n/a | 0.000 | 1984 | 1000 | NONE |
| R-407B | R-32/125/134a (10/70/20) [Klea61] | n/a | 0.000 | 2692 | 1000 | NONE |
| R-407C | R-32/125/134a ( 23/25/52) [Klea66] | H M | 0.000 | 1640 | 1000 | NONE |
| R-507 | R-125/143a (50/50) [AZ-50] | n/a | 0.000 | 3600 | 1000 | NONE |
| R-600 | $CH_3CH_2CH_2CH_3$ [butane] | H M L | 0.000 | 3 | 800 | 1.5 |
| R-717 | $NH_3$ [ammonia] | H M L | 0.000 | neglig | 25 | 15 |
| R-718 | $H_2O$ [water] | n/a | 0.000 | 0 | 0 | NONE |
| R-290 | $CH_3CH_2CH_3$ [propane] | H M L | 0.000 | 3 | 1000 | 2.1 |

*Sources: UNEP. 1994. Report of the Refrigerant, Air Conditioning and Heat Pumps Technical Options Committee. Draft.; Bitzer Kuhlmaschinenbau. 1994. Bitzer Refrigerant Report 2.*
*Application range: HH Extra high temp; H High temp; M Medium temp; L Low temp; Appl. range indicated in brackets is not preferred
**Lower flammability limits are expressed as volume % in ambient air.
†Some refrigerant names and/or synonyms are manufacturer's trade names.

*FIGURE 1. Refrigerant properties.*

stability of oil and refrigerant, leading to possible formation of acids that can damage copper plating.

**Safety and Health:** R-22 is classified as an A1 refrigerant, meaning it is nonflammable and nontoxic. However, R-22 is combustible at pressures above atmospheric in the presence of high air concentrations. As with many refrigerants, R-22 is heavier than air and can cause asphyxiation if released in sufficient quantities. The revision of ASHRAE Standard 15-92 will require a refrigerant sensor/monitor to be installed in all mechanical rooms for all refrigerants.

## R-123

**Process:** HCFC-123 (also known as SUVA 123) is currently limited to use in centrifugal chillers. It has been identified as a primary replacement for R-11 in new chillers and in most existing low-pressure chillers. It is currently the only retrofit for R-11 chillers.

**ODP:** 0.020

**Direct GWP:** 90

**Materials Compatibility:** R-123 is a more aggressive solvent than R-11, leading to concerns that R-123 may not be compatible with current hermetic motor insulations. Plastics and elastomers may have to be replaced with materials that are compatible with R-123.

Testing shows R-123 to be miscible with existing refrigeration lubricants, such as alkylbenzene, paraffinic, and napthenic lubricants.

**Safety and Health:** R-123 is classified as a B1 refrigerant and a refrigerant vapor detector with a range of 0 ppm to 150 ppm is required to be placed in the machinery room and directed mechanical ventilation or localized exhaust should be used to maintain airborne concentrations at safe levels not requiring respirator protection. R-123's maximum working pressure is low enough to exempt the systems from having to meet pressure vessel code requirements. It is not flammable. EPA does recommend that users adhere to requirements of ASHRAE Standards 15 and 34. When handling R-123, appropriate personal protective equipment should be worn. These include eye protection, gloves, and safety shoes. In the event of a leak, respirators should be available for immediate use and individuals who work with R-123 regularly should be trained in the proper use of respirators, as well as, fit-tested annually to ensure adequate fit and proper function. Coverall chemical goggles and a face shield should be used when making first breaks into a system or drum if liquid splash is a potential problem.

## BLENDS CONTAINING R-22

There are currently several blends under development that use HCFC-22 as the primary component. These are being considered mainly for retrofit of systems using R-11 and R-502 where R-22 is not considered an optimal retrofit alternative. Use of these blends may offer a simpler and less expensive retrofit of these systems in some cases. The addition of chlorine-free substances significantly reduces R-22's high discharge-gas temperature and increases the applications for which these blends can be used. One of the disadvantages of these blends is that, because they contain R-22 as a primary component, they will not be able to be manufactured after 2020. Additionally, their relatively high cost, and concerns over whether they will fractionate into their component parts, are issues to be considered.

## R-401 A and B

Process: R-401A (also referred to by its brand name MP39) is a blend of R-22/152a/124 in a 53/13/34 percent composition. R-401B (referred to by its brand name MP66) is a blend of the same refrigerants in a 61/11/28 percent composition. The 401 series is being considered as a replacement for R-12 and R-500 in medium-pressure applications.

**ODP:** R-401A has an ODP of 0.037. R-401B has an ODP of 0.040.

**Direct GWP:** The GWP of R-401A is 1018* and the GWP of R-401B is 1116.

**Materials Compatibility:** The manufacturer does not recommend that these refrigerants be used in systems with flooded evaporators as they may affect performance. These blends are not miscible with mineral oils sometimes used in R-12 systems. In general, elastomers recommended for use with R-22 will be compatible with these refrigerants, however, the equipment manufacturer should be consulted prior to retrofitting.

**Safety and Health:** This refrigerant has not been classified by ASHRAE but, the manufacturer indicates that this refrigerant is non-flammable and has a low toxicity. However, these refrigerants should not be exposed to open flame or electrical heating elements. These refrigerant blends present no acute or chronic hazard when handled in accordance with recommendations and when exposures are maintained at or below the recommended exposure limits. Inhaling high concentrations of these refrigerants may cause temporary central nervous system depression with narcosis (sleepiness), lethargy, and weakness. It may also cause dizziness, intoxication, and loss of coordination. Continued breathing may produce cardiac irregularities, unconsciousness, or death. Use self-contained breathing apparatus, when necessary, to avoid exposure to fumes. Avoid exposure to skin as refrigerants can cause dryness, irritation, and, for low- to medium-temperature refrigerants, frostbite. Decom-

---

*Global warming potentials for blends were calculated by adding the GWPs of the component parts based on their percentage in the blend.

position may produce toxic and irritating compounds such as hydrogen chloride and hydrogen fluoride.

## R-402 A and B

**Process:** R-402 A (also referred to under its brand name HP80) is a blend of R-125/290/22 in a 60/2/38 percent ratio. R-402 B (referred to under the brand name HP81) is a blend of the same refrigerants in a 38/2/60 percent ratio. These blends are being considered for retrofit of equipment using R-502. Some retrofits using HP80 have already been performed on supermarket applications.

**ODP:** The ODP of R-402A is 0.021. The ODP of R-402B is 0.033.

**Direct GWP:** The GWP of R-402A is 2648 and the GWP of R-402B is 2252.

**Materials Compatibility:** R-402 is immiscible in lubricants generally used with R-502. R-402 will work with either alkylbenzene lubricants or polyol esters.

**Safety and Health:** ASHRAE has rated these refrigerant as A1. They are nonflammable with a low degree of toxicity. R-402A and R-402B require the same safe handling procedures as R-502. Inhaling high concentrations of these refrigerants may cause temporary central nervous system depression causing dizziness, and loss of coordination. Continued breathing may produce cardiac irregularities, unconsciousness, or death. Use self-contained breathing apparatus, when necessary, to avoid exposure to fumes. Avoid exposure to skin as refrigerants can cause dryness, irritation, and frostbite. Decomposition may produce toxic and irritating compounds such as hydrogen chloride and hydrogen fluoride.

## HYDROFLUOROCARBONS (HFCs)

The HFCs are considered primary replacements for most current CFC, and some HCFC, uses. These compounds have proven to possess properties similar to those of CFCs and are more stable than HCFCs (except for HCFC-22). The absence of chlorine renders these compounds nonreactive with stratospheric ozone and, therefore, less damaging to the ozone layer. The tradeoff is that these compounds tend to have very long atmospheric lifetimes, which increases their ability to contribute to global warming.

## R-134A

**Process:** R-134a is considered to be the leading candidate for replacement of R-12 in many applications. R-134a is also used in centrifugal chillers from approximately 350 kW to 4500 kW capacity, and is being considered as a replacement for R-22 in such large chillers. R-134a may also be viable as a replacement for R-22 in unitary equipment.

**ODP:** 0

**GWP:** 1200

**Materials Compatibility:** The greatest material compatibility concern associated with R-134a is in relation to lubrication. R-134a is immiscible in most commonly used refrigerant lubricants. Polyalkylglycol lubricants have been developed for use with R-134a, but these lubricants are not compatible with R-12 residues. This means that R-12 systems need to be extensively flushed prior to retrofitting with R-134a. Polyolester oils are now widely used with R-134a and have overcome some of the compatibility issues. However, several flushings of the system are generally recommended. Testing is currently underway on the compatibility of R-134a with common equipment construction materials such as copper, steel, and aluminum. Some elastomers and plastics currently used in existing equipment may also be incompatible with R-134a. Gaskets, shaft seals, and o-rings should be reviewed with the equipment manufacturer before retrofit.

**Safety and Health:** ASHRAE has provisionally rated this refrigerant as A1, pending completion of toxicological testing. R-134a is not flammable at ambient temperatures and atmospheric pressure. However, tests have shown it to be combustible at pressures as low as 5.5 psig at 177 degrees C (350 degrees F) when mixed with air at concentrations of generally more than 60 percent. Therefore, it should not be mixed with air for leak testing. At lower temperatures, higher pressures are required for combustibility. R-134a presents no acute or chronic hazard when handled in accordance with recommendations and when exposures are maintained at or below the recommended exposure limits. Inhalation of high concentrations of R-134a may cause temporary central nervous system depression, with narcosis (sleepiness), lethargy, weakness, dizziness, intoxication, and loss of coordination. Continued breathing may produce cardiac irregularities, unconsciousness, or death. Use self-contained breathing apparatus when necessary to avoid exposure to fumes. Avoid exposure to skin as R-134a can cause dryness, irritation, and, for low to medium temperature refrigerants, frostbite. Decomposition of R-134a may produce toxic and irritating compounds such as hydrogen chloride and hydrogen fluoride.

## BLENDS CONTAINING HFCs

Refrigerant blends containing HFCs have been developed as long-term replacements for applications using R-502 and are currently on the market. HFC blends for replacing R-22 are currently under development. The need to find replacements for R-22, which is facing a production phaseout in 2020 or even sooner, has focused attention on the HFC blends that can perform as well as, and in some cases better than, R-22 in certain applications. However, there is no single alternative that stands out as a universal replacement for either R-22 or R-502. These blends have the advantage of being free of chlorine, and therefore having no ozone-depletion potential. The relatively high direct global-warming potential is a possible issue; however, the increased energy efficiency of these blends may lead to an overall lower TEWI value[3].

## R-404A

**Process:** R-404a (also known as HP62) is a blend of R-125/143a/134a in a 44/52/4 percentage ratio. It is being developed as a replacement for R-502.

**ODP:** 0

**Direct GWP:** 3520

**Materials Compatibility:** Because R-404 contains R-134a as a primary component, it should have similar materials compatibility concerns. These include immiscibility with commonly used lubricants; compatibility concerns with equipment construction materials, such as copper, steel, and aluminum, and potential incompatibility with certain plastics and elastomers. R-404a should be used with polyol ester lubricants only and should never be mixed with other refrigerants as this may cause system damage. Gaskets, shaft seals, and o-rings should be reviewed for compatibility before retrofitting with R-404a. This refrigerant will maintain consistent composition if leakage occurs. Operating performance will remain consistent following leakage/recharge cycles.

**Safety and Health:** ASHRAE has designated this refrigerant as A1 meaning it is of low toxicity and nonflammable. R-404a can be hazardous if used improperly. These hazards include pressure and frostbite from escaping refrigerant. Prolonged exposure to high concentrations can cause asphyxiation and cardiac sensitization (irregular heartbeat).

## R-407 A, B, AND C

**Process:** R-407 A (also referred to under its brand name KLEA 60) is a blend of R-32/125/134a in a 20/40/40 percent ratio and is an acceptable substitute for R-502 in new refrigeration equipment and for retrofit in many existing systems using R-502. R-407 B (also known as KLEA 61) is a blend of R-32/125/134a in a 10/70/20 percent ratio. It is designed to replace R-502 in existing applications in which R-407 A is not appropriate, such as hermetic compressors where discharge temperature is a critical characteristic. These blends are also being considered for replacement of R-22. R-407 C (also known under the trade name of KLEA 66) is the ASHRAE designation (pending approval) for a refrigerant blend of R-32/125/134a in a 23/25/52 percent ratio. It is designed to replace HCFC-22 in new low-temperature refrigeration equipment and for retrofit in many existing systems.

**ODP:** 0

**Direct** GWP: The GWP of R-407 A is 1984; the GWP of R-407 B is 2692; the GWP of R-407 C is 1640 .

**Materials Compatibility:** R-407 A and B should be used with synthetic neopentyl polyolester lubricants. Tests have shown that these lubricants are stable in the presence of R-407 when in contact with copper, steel, or aluminum.

**Safety and Health:** Although R-407 contains the marginally flammable refrigerant R-32, its compositions have been designed so that they are nonflammable and exhibit low toxicity.

## NATURAL REFRIGERANTS

The primary natural refrigerants in common usage today include water, ammonia, lithium bromide, nitrogen, and carbon dioxide. Ammonia, in particular, has long been used as a refrigerant in many types of applications. The natural refrigerants are gaining new attention because of the CFC and eventual HCFC phaseout, and because concerns over global warming increasingly call into question the long-term future of the HFCs. Whether the natural refrigerants will be able to take the place of CFCs and HCFCs in a range of applications, however, remains questionable. The natural refrigerants are gaining increased acceptance in Europe, where the phaseout schedule for HCFCs is expected to be significantly accelerated.

### R-717 (Ammonia)

**Process:** Ammonia has been used in a wide variety of refrigerant applications for many years. It is a very efficient refrigerant because of its heat-transfer properties and its low molecular weight. Its most prevalent application is in food-processing and storage applications. Ammonia is being considered as a replacement for CFCs and HCFCs in industrial applications, and in some HVAC applications where it is used as a secondary refrigerant (absorption systems). It is also used as a primary refrigerant to chill a secondary working fluid such as water or brine for circulation.

**ODP:** 0

**GWP:** negligible

**Materials Compatibility:** Ammonia is not compatible with copper. In the presence of moisture (water vapor), ammonia forms highly corrosive acids (nitric & nitrous acids) that can vigorously attack some metals, particularly copper-based alloy components such as motor windings. Significant research is being undertaken to identify suitable lubricants for use with ammonia. Previously used oils were not soluble with the refrigerant.

**Safety and Health:** Ammonia is classified as a B2 refrigerant (of moderate toxicity and flammability). Although it is not lethal in concentrations below 2500 ppm, it has a threshold limit of 25 ppm. Ammonia gives off flammable vapors which become explosive when exposed to air. Closed containers exposed to heat may explode. Inhalation of vapors may cause severe irritation or burns of the respiratory system, pulmonary edema, or lung inflammation. Contact with skin or eyes may cause severe irritation or burns. Prolonged eye contact may cause permanent damage to the cornea and blindness may occur. Existing OSHA (Occupational Safety and Health Administration) and ASHRAE standards are sufficient to reduce the risks associated with ammonia use. However, users should also check local building codes.

### R-718 (Water)

**Process:** Water has long been used as a refrigerant in absorption systems. It is also used extensively in evaporative and desiccant cooling applications. Because of its high vapor pressure, it is not considered an efficient refrigerant for use in vapor compression cycles. It would require a very large or very high speed compressor to be effectively used in a vapor compression cycle.

**ODP:** 0

**GWP:** 0

**Materials Compatability:** Water is a potent solvent and can degrade metals. Water becomes corrosive in the presence of oxygen. Water is considered a non-regulated product, but may react vigorously with some specific materials. Avoid contact with all materials until investigation shows substance is compatible.

**Safety and Health:** R-718 is classified as an A1 refrigerant. It is neither toxic nor flammable. In absorption applications it is used in concert with either lithium bromide or ammonia. Safety considerations associated with these compounds should be considered.

## HYDROCARBONS

Hydrocarbons are long-term, proven refrigerants with excellent refrigerant properties. They do not contain chlorine or bromine and thus have no ODP. However, they do degrade in the lower atmosphere and contribute to air pollution. Propane, ethane, propylene, and butane are currently used as refrigerants in specialized industrial applications, such as those found in oil refineries and chemical plants that handle other flammable fluids. The major drawbacks of hydrocarbon refrigerants is that they are highly flammable at low concentrations and generally have very low ignition temperatures compared with other refrigerants.

## HC-290 (Propane)

**Process:** Propane is an acceptable substitute for CFC refrigerants R-11, R-12, and R-502 in industrial process refrigeration. It is being considered for use in applications having a very low refrigerant charge (such as domestic refrigerators) in Europe.

**ODP:** 0

**GWP:** 3

**Materials Compatibility:** Incompatible with strong oxidizing materials. Avoid contact with heat, spark, flame, peroxides, plastics, and chlorine dioxide. Polymerization will not occur.

**Safety and Health:** Propane is classified as an A3 refrigerant (highly flammable but of low toxicity). EPA recommends that it only be used at industrial facilities that manufacture or use hydrocarbons in the process stream. Such facilities are designed to comply with the safety standards required for managing flammable chemicals. Overexposure can cause weakness, headache, confusion, blurred vision, nausea, drowsiness, and other nervous system effects. Greater exposure may cause dizziness, slurred speech, flushed face; eye irritation, skin and lung irritation, unconsciousness, convulsions, and asphyxiation. High concentrations (>10%) may lead to death. Contact with the liquid product can cause frostbite. The odorant can irritate the eyes, skin, and respiratory tract.

## R-600 (Butane)

**Process:** Butane is an acceptable substitute for CFC refrigerants R-11, R-12, and R-502 in industrial process refrigeration.

**ODP:** 0

**GWP:** 3

**Materials Compatibility:** Incompatible with strong oxidizing agents. Will not polymerize.

**Safety and Health:** Butane is classified as an A3 refrigerant (highly flammable but of low toxicity). EPA recommends that it only be used at industrial facilities that manufacture or use hydrocarbons in the process stream. Such facilities are designed to comply with the safety standards required for managing flammable chemicals. Data are insufficient to indicate effects of exposure.

## CHLOROFLUOROCARBONS

For comparison with the refrigerants discussed above, the properties of the CFC refrigerants are provided below.

## R-11

**Process:** R-11 is a low-pressure refrigerant commonly used in centrifugal and screw commercial chilled-water air conditioning applications. R-11 has also been used in medium-sized, low-pressure centrifugal chillers for making cold process water.

**ODP:** 1.0

**GWP:** 3400

**Materials Compatibility:** R-11 is incompatible with alkali or alkaline earth metals (powdered aluminum, zinc etc.). Polymerization will not occur.

**Safety and Health:** R-11 is classified as an A1 refrigerant. Inhalation of vapor is harmful and may cause heart irregularities, unconsciousness or death. Overexposure may cause eye irritation and skin rash or discomfort. Overexposure by inhalation may cause temporary nervous system depression with dizziness, headache, confusion, incoordination, and loss of consciousness; temporary heart irregularities including irregular pulse and palpitation may occur as well as death with grossly excessive exposures. Individuals with diseases of the central nervous system or cardiovascular system may have increased susceptibility to the toxicity of excessive exposures. Decomposition products generated through exposure of chemical to high temperatures are hazardous.

## R-12

**Process:** R-12 is a medium-pressure refrigerant commonly used in centrifugal and screw commercial chilled-water air conditioning applications. R-12 has also been used as a medium temperature refrigerant (>-30C) in industrial and commercial applications, especially for cold storage.

**ODP:** 1.0

**GWP:** 7100

Materials Compatibility: R-12 is incompatible with alkali or alkaline earth metals (powdered aluminum, zinc etc.). Polymerization will not occur.

**Safety and Health:** R-12 is classified as an A1 refrigerant. Inhalation of high concentrations of vapor is harmful and may cause heart irregularities, unconsciousness or death. Contact with the vapor may cause eye irritation, tearing, and blurring of vision. Skin contact may cause temporary nervous system depression, with dizziness, headache, confusion, incoordination, and loss of consciousness; temporary heart irregularities, including irregular pulse and palpitation, may occur as well as death with grossly excessive exposures. Individuals with diseases of the central nervous system

or cardiovascular system may have increased susceptibility to the toxicity of excessive exposures. Decomposition products generated through exposure of chemical to high temperatures are hazardous.

## R-114

**Process:** R-114 is a low-pressure refrigerant used extensively in centrifugal compressors to provide chilled-water for a variety of HVAC applications as well as some very large specialized industrial process cooling. Used extensively in naval shipboard applications.

**ODP:** 1.0

**Direct GWP:** 7000

Materials Compatibility: R-114 is incompatible with alkali or alkaline earth metals (powdered aluminum, zinc, etc.). Polymerization will not occur.

**Safety and Health:** ASHRAE lists this as an A1 refrigerant. Inhalation of high concentrations of vapor is harmful and may cause heart irregularities, unconsciousness, or death. Liquid contact may cause frostbite. Overexposure may cause eye irritation and skin rash. Individuals with diseases of the central nervous system or cardiovascular system may have increased susceptibility to the toxicity of excessive exposures. Decomposition products generated through exposure of the chemical to high temperatures are hazardous.

## R-500

**Process:** R-500 is a high-pressure refrigerant commonly used in centrifugal and screw commercial chilled-water air conditioning applications. R-500 has also been used as a medium-temperature refrigerant (>-30°C) in industrial and commercial applications, especially for cold storage.

**ODP:** 0.738

**Direct GWP:** 5254

**Materials Compatibility:** R-500 is incompatible with alkali or alkaline earth metals (powdered aluminum, zinc, etc.). Polymerization will not occur.

**Safety and Health:** R-500 is classified as an

A1 refrigerant. Eye contact may cause discomfort, tearing, or blurring of vision. Inhalation may cause nausea, headache, or weakness; or confusion and lack of consciousness. Inhalation of high concentrations (>20%) of vapor, and skin contact are harmful and may cause difficulty breathing, irregular pulse, heart irregularities, kidney irregularities, unconsciousness or death. Liquid contact may cause frostbite. Individuals with diseases of the central nervous system, lungs, kidneys, or cardiovascular system may have increased susceptibility to the toxicity of excessive exposures. Decomposition products generated through exposure of the chemical to high temperatures are hazardous.

## R-502

**Process:** has been used as a medium-temperature refrigerant (>-45°C) in industrial and commercial applications.

**ODP:** 0.283

**Direct GWP:** 5534

**Materials Compatibility:** R-502 is incompatible with alkali or alkaline earth metals (powdered aluminum, zinc, etc.). Polymerization will not occur.

**Safety and Health:** R-502 is classified as an A1 refrigerant. Inhalation of high concentrations of vapor is harmful and may cause nausea, headache, weakness, irregular pulse, heart irregularities, unconsciousness, or death. Eye contact may cause discomfort, tearing, or blurring of vision. Liquid contact may cause frostbite. Individuals with diseases of the central nervous system, lungs, kidneys, or cardiovascular system may have increased susceptibility to the toxicity of excessive exposures. Decomposition products generated through exposure of the chemical to high temperatures are hazardous.

## CONCLUSIONS

This chapter has provided some general information on the currently available replacement refrigerants. The following chapter will provide a discussion of which refrigerant options are viable as retrofit alternatives, and which are being identified as replacements.

## REFERENCES

1. American Society of Heating, Refrigerating and Air-Conditioning Engineers, ASHRAE Standard 34-1992, ASHRAE, Atlanta, Georgia.

2. United Nations Environment Programme, Montreal Protocol 1994/5 Assessment, Refrigeration, Air Conditioning and Heat Pumps Technical Options Committee, Draft Chapter 7, *Air Conditioning & Heat Pumps,* July 1994. Nairobi, Kenya.

3. Bitzer Kühlmaschinenbau, Refrigerant Report 2, No. 9306 E, D-71044 Sindelfingen, Germany.

# CHAPTER 10

# ALTERNATIVE ROUTES
## (Applications for Alternatives)

The previous chapter described various refrigerants being identified as replacements for currently used halocarbon refrigerants. Your eventual goal, of course, is to make the transition to environmentally friendly alternative refrigerants or refrigeration processes. This chapter discusses the choices available for equipment retrofits and replacement, including replacement with not-in-kind technologies. This information is also shown in Figure 1.

### RETROFITS

A number of the refrigerants discussed in Chapter 9 have been developed specifically as interim retrofits for use in existing equipment. As mentioned earlier, so far there is no "silver bullet" replacement that can be put into the system with no other changes required. The amount of change that needs to be made to the equipment will depend on the type of alternative refrigerant chosen and the type of equip-

| Refrigerant | Transitional Chemical Substitute | Chlorine-free Chemical Substitute | Alternative Technologies |
|---|---|---|---|
| CFC-11 | HCFC-123, HCFC-22, HCFC-124 | HFC-134a,[*] R-290, R-600, R-717 | Water/lithium bromide absorption Ammonia/water absorption Evaporative cooling Desiccant cooling |
| CFC-12 | HCFC-22, HCFC-123, HCFC-124, R-401A,[**] R-401B, R-402A, R-402B | HFC-134a R-404, R-507, R-290, R-600, R-717 | Water/lithium bromide absorption Ammonia/water absorption Evaporative cooling Desiccant cooling Carbon dioxide ($CO_2$) |
| CFC-500 | HCFC-22, HCFC-123, HCFC-124, R-401A, R-401B, R-402A, R-402B | HFC-134a, R-404A, R-507, R-290, R-600 , R-717 | Water/lithium bromide absorption Ammonia/water absorption Evaporative cooling Desiccant cooling |
| CFC-502 | HCFC-22, R-401A, R-401B, R-402A, R-402B | HFC-134a, R-404A, R-507, R-407A, R-407B, R-290, R-600, R-717 | Evaporative cooling Desiccant cooling |
| HCFC-22 | | HFC-134a, R-407C, R-717, R-290, | |

[*] Use of HFCs is subject to the no venting prohibition under Section 608(c)(2), which takes effect November 15, 1995, at the latest.

[**] The R-400 series refrigerants and HCFC blends are subject to containment and recovery regulations covering HCFCs

*FIGURE 1. Refrigerant alternatives.*

ment in which it is used. In general, flammable and/or highly toxic refrigerants are ruled out as retrofits because of the extensive changes required for the systems to comply with strict safety standards. Some applications will not have a viable retrofit. Some losses in the system capacity and energy efficiency can be anticipated with most retrofits. However, custom engineering of the retrofit may be able to offset some of the these losses.

## R-123

R-123 was specifically designed to be used as a retrofit for R-11 centrifugal chillers. Several conversions from R-11 to R-123 have already been successfully accomplished. R-123 is a relatively efficient refrigerant; however, some losses in efficiency and capacity can be anticipated in a straight retrofit where no significant changes are made to the equipment. Depending on the size and application of the equipment, efficiency losses have been predicted to range from 1%-10%[1], and anticipated capacity losses can range from 5% to 20%[2]. Efficiency losses are primarily due to the design of the impeller and to the evaporator and condenser tube surfaces, which were originally optimized for R-11. In general, the larger capacity losses occur in larger machines with higher impeller speeds. An engineered retrofit can help to alleviate capacity and energy efficiency losses.

In addition to capacity and efficiency considerations, retrofitting equipment to use R-123 must also take into account the potential need to replace elastomers and plastic components of the R-11 equipment, which may be incompatible with R-123. R-123 also has a lower allowable exposure level (AEL) threshold value than R-11. Changes to the mechanical room may be required to ensure adequate protection of technicians and others who may come in contact with the refrigerant. The types of changes vary depending on the application, with industrial, unoccupied spaces requiring the lowest level of regulation.

## HFC-134a

Several retrofits of systems using R-12 and R-500 to R-134a have been successfully accomplished. HFC-134a has been identified as the primary retrofit for R-12 and R-500 in existing equipment for applications having evaporator temperatures of -7°C or higher. At lower temperatures, use of R-134a will generally result in lower capacity compared to R-12, and thus may require modification to the compressor. Even at evaporator temperatures of −7°C, the retrofit may require the impeller speed in R-12 equipment to be increased 10% to 15%, or the impeller may have to be replaced with one designed for use with R-134a. Capacity changes from R-134a versus R-12 range from 1% to 9%, with efficiency losses ranging from 1% to 10%.

In evaluating the feasibility of a retrofit, it is important to keep in mind that R-134a is not compatible with the lubricants used in R-12 systems, and therefore requires several system flushings to remove existing lubricant concentrations to below 10% for simple chilled-water applications and 1% to 2% for low-temperature applications[3]. Systems are routinely flushed with R-11, which is then recovered and used in R-11 equipment. In addition, R-134a may be incompatible with other components of the existing equipment, including copper, steel, aluminum, and some elastomers and plastics.

## R-401 A/B

R-401a (MP39) and 401b (MP66) are being considered as retrofits for R-12 and R-500 in direct expansion systems using positive displacement compressors and operating in medium- to low-temperature ranges. They are not recommended for use in systems with flooded evaporators because of the difference in vapor composition and liquid composition of these refrigerants. These blends, which contain R-22 as a primary component, are being considered in lieu of a retrofit to R-22, because of the higher volumetric capacity of R-22 compared with R-12. Some conversions using these refrigerants have been accomplished. R-401a is recommended for retrofitting of medium-temperature R-12 systems operating at evaporator temperatures of -23C and above. R-401b is suitable for R-12 systems operating at evaporator temperatures below -23C, and for R-500 systems. Capacities and efficiencies will be similar to those of the refrigerants being replaced.

Both R-401a and R-401b are incompatible with the mineral oil lubricants sometimes used in R-12 systems. Alkylbenzene lubricants, which may be used in current R-12 and R-500 systems, are compatible with R-401 A/B. Plastics and elastomers compatible with R-22 should also be compatible with R-401 A/B. Retrofits of R-12 systems generally require a slightly smaller charge of R-401A/B than of R-12, while R-500 equipment generally requires a slightly larger charge. Additional adjustments to the equipment to account for the different thermodynamic properties of the replacements may also be required.

## R-402A/B

R-402A (HP80) and R-402B (HP81) were designed as retrofits for equipment using R-502. R-402a has been used successfully in the retrofit of several supermarket and restaurant applications. R-402 A/B, which contains R-22 as a primary component, tends to perform better than R-22 in the lower temperature ranges. R-402 has properties similar to the R-502 that it is replacing, keeping efficiency and capacity losses to a minimum.

R-402 is not compatible with lubricants commonly used in R-502 applications, requiring a change in lubricants when retrofitting. R-402 is compatible with both alkylbenzene and polyol ester lubricants. R-402 has been developed to require minimal system changes during retrofit.

## REPLACEMENT

In general, there are more options available to the equipment owner who plans to replace, rather than retrofit his or her existing equipment. The choices range from a number of manufactured alternative refrigerants, to not-in-kind technology changes. The choice of whether to stay with a given technology and replace the existing equipment with equipment that utilizes an alternative refrigerant, or to change to an alternative technology, will depend on the type of application, the geographic location, and any building and safety code requirements. Space-cooling applications for use in occupied spaces often have strict safety and health requirements that can preclude use of certain refrigerants or technologies.

In addition to the benefits from changing out of CFCs, one advantage to replacing existing equipment is the potential energy-efficiency gains. By using equipment that was specifically designed for the refrigerant, you are more likely to realize energy gains than through a retrofit. In addition, some alternative technologies can offer their own energy-efficiency advantages.

## R-22

R-22 is not generally considered a new refrigerant, since it has been used extensively in air-cooled air-conditioning and heat-pump applications for many years. When the initial phaseout of CFC refrigerants was announced, R-22 quickly became the favorite candidate for replacement of R-12, R-500, and R-502 in many industrial applications, and as the working fluid for high-pressure, positive-displacement chillers. Although the dominance of R-22 is being challenged in some applications by the development of other chemical alternatives, it still remains a strong candidate for many uses in the United States.

Because it is an HCFC, production of R-22 is scheduled to be phased out by 2020—this is earlier than the schedule for most other HCFCs because of the high ozone-depletion potential of this refrigerant. In Europe, where the use of alternatives such as hydrocarbons and ammonia is not as strictly regulated, there is discussion of accelerating the production phaseout of R-22 and other HCFCs to the end of this century. The phaseout schedule for the HCFCs has already undergone one acceleration at the Copenhagen meeting of the *Montreal Protocol* parties in 1992. If R-22 will continue to be produced until 2020, the fact that it is being phased out should not unduly affect decisions to purchase new equipment at this time, as recycled R-22 will likely be available for some time after the phaseout, and the time between now and the phaseout should be adequate for equipment owners to recognize a return on investment.

The higher discharge temperatures associated with R-22 use may require replacing single-stage R-502 equipment with two-stage R-22 equipment. This would result in an increase in cost of the replacement equipment over replacement with single-stage systems.

## R-123

R-123 was mainly designed as a retrofit chemical for use in existing R-11 chillers. However, it is also being used a refrigerant in new chillers being designed in the United States. It has been successfully used in a new series of centrifugal compressors that have provided energy efficiencies as high as .62 kW/ton. These are some of the lowest available at this time and are very near the theoretical limits.

Because it is an HCFC, production of R-123 is scheduled to be phased out by 2030. As mentioned above, this time frame should not preclude use of R-123 in new equipment. The phaseout schedule is also not anticipated to affect the supply or price of this refrigerant.

## R-134a

R-134a is currently being used in positive-displacement systems such as water chillers as a replacement for R-12. Because the flow characteristics of R-134a and R-12 are similar, the compressor and equipment sizes are also similar. Thus, the cost of an R-134a replacement chiller is similar to the current cost for R-12 equipment. The theoretical efficiency of R-134a is approximately 2% lower than R-12, but this is generally not a problem in new equipment due to changes in the efficiency of other equipment components.

R-134a is gaining a share of the industrial market, especially in applications operating above the -10C temperature ranges. At these temperatures, cooling capacity and efficiency are not substantially different from R-12. At lower temperatures, a 15% increase in energy consumption and a 30% loss in capacity may result. Although these losses may be compensated for through equipment design (using extensive liquid subcooling), higher-pressure refrigerants are usually preferred for these applications.

R-134a is also being examined as a replacement for R-22 in large air-cooled chillers. It is possible to design unitary equipment that uses R-134a and achieves similar efficiency and capacity as R-22. However, this equipment generally has larger heat-exchanger and refrigerant tubing, larger compressors, and re-sized compressor motors. These changes all increase the expense and size of the equipment. It is estimated that for a coefficient of performance of 4.0 or greater, equipment may be 30% to 40% more expensive than current R-22 equipment. As cooling-efficiency targets are raised even higher, R-134a equipment may become impractical to construct.[4]

## AMMONIA

Ammonia is currently a primary refrigerant for use in many industrial applications, especially cold-storage applications for which, it has been estimated, as much as 80% of the refrigerated warehouses in the United States operate on ammonia. Ammonia is a highly effective refrigerant, more so even than the halocarbons, because of its excellent heat-transfer properties and low molecular weight. New developments in lubrication have led to dramatic improvements in the efficiency of ammonia equipment. Ammonia can be used in either vapor-compression equipment, or in absorption systems where it operates as the absorbent.

Ammonia is gaining significant attention in Europe, where the accelerated phaseout scheduled for HCFCs, and potential global-warming controls on HFCs, are limiting other refrigerant alternatives. Implementation of ammonia technology in Europe has been helped by the introduction of semithermetic ammonia compressors and soluble lubricants. Ammonia systems have generally been manufactured only with open drives because of the incompatibility of ammonia with the copper windings on motors. Research is also underway to develop special evaporators that can reduce the overall refrigerant charge, thereby alleviating some of the safety concerns.

In the United States, concerns over ammonia's flammability and toxicity have tended to limit the applications in which it has been used. Current regulations in the United States limit ammonia chillers to large systems that are isolated from the general public. Many concerns regarding ammonia's toxicity can be overcome by proper system design and good management practices. Improvements in system design have significantly reduced the possible hazards of ammonia in the event of a system breakdown. The very strong odor associated with ammonia (at concentrations as low as 10 ppm) not only makes it "self-alarming" but simplifies leak detection.

Although ammonia has been used extensively in industrial applications for over 100 years, it will have difficulty gaining acceptance in commercial applications for occupied spaces because of the strict regulatory constraints. ASHRAE Standard 15 limits the use of ammonia in public buildings to those systems that utilize a secondary heat-transfer fluid, while confining the ammonia itself to the machine room. Ammonia absorption systems, which may be used in certain situations for air conditioning applications, are significantly less effective than vapor-compression systems. In fact, the absorption systems have efficiencies of only 50% to 60% of the compressor systems. A primary advantage to single-effect absorption systems is their ability to operate on waste heat generated through other processes. In addition, in places where natural gas is available and compares favorably to electricity, the use of natural gas can reduce the efficiency losses associated with absorption systems.

## WATER

As discussed in the section dealing with ammonia, water is used as the refrigerant in some absorption systems. Water is also used as a coolant in passive cooling or evaporative and desiccant cooling systems. Evaporative systems have traditionally been used for air conditioning in hot, arid areas with low relative humidity. Evaporative systems that use water as a coolant can initially be more expensive than traditional CFC- and HCFC-based systems. However, energy efficiency is usually increased so much with evaporative systems, that for some scenarios the payback can be recognized in as little as one year.

## DESICCANTS

The use of desiccant systems in conjunction with evaporative cooling processes has expanded the geographic areas where the evaporative systems can be applied by decreasing the impact of humidity on the system performance and theoretical temperature limits. As discussed in Chapter 2, desiccant-based systems have the advantage of reducing humidity and improving indoor air quality while cooling the air. Moreover, desiccants systems can reduce moisture in the air at higher temperatures than normal vapor-compression systems. Thus, areas with tight humidity controls do not have to be "overcooled" just to reduce humidity. Since these systems tend to be large, they are primarily used in commercial applications. In particular, desiccant-based cooling has been gaining popularity as a replacement at supermarkets, hotels, ice

rinks, hospital operating rooms, and commercial buildings.

## HYDROCARBONS

Hydrocarbon refrigerants are also gaining attention as replacements for CFCs and HCFCs in Europe. As with the interest in ammonia, this is spurred by the faster phaseout scheduled for HCFCs, and potential global-warming controls on HFCs, which are limiting other refrigerant alternatives. Most of the hydrocarbon evaluation has been done on R-290 (propane). R-290 has refrigeration capacity and pressure levels that are similar to R-22 and R-502, and temperature properties that are similar to R-12 and R-502. Hydrocarbons do not have the materials compatibility problems associated with ammonia, and are generally compatible with most mineral oils currently in use in CFC systems.

R-290 is a proven refrigerant, with refrigeration plants using R-290 having been in operation world-wide for many years. Significant research is currently being undertaken on the use of R-290 in compact systems for air conditioning, transport refrigeration and heat pumps. R-290, when blended with R-600a (isobutane), is being investigated as a replacement for R-12 in small operations (specifically domestic refrigerators).

The primary disadvantage to hydrocarbons is their flammability. This severely restricts their use in applications in the United States. In general, hydrocarbon use in the U.S. is restricted to unoccupied areas or to equipment with very low refrigerant charge levels. In industrial applications, hydrocarbons provide a viable alternative for halocarbon refrigerants because of their extreme flexibility (they can be used in any temperature range), their excellent refrigerant properties, and their low environmental impacts. Hydrocarbon refrigerants have been used in the oil and gas industry, but have generally not found their way into other industrial applications, even where other flammable chemicals are used.

## CONCLUSIONS

This chapter has provided an overview of some of the issues associated with alternative refrigerants used as retrofits or replacement for existing systems. To a large extent, the capacity and energy efficiency issues associated with

ments will be based on site-specific and application-specific factors. The following chapters will provide a discussion of the issues that need to be considered in determining whether to retrofit or replace your existing equipment. In addition, once you have made some initial judgements, you should work with the company that will be undertaking the retrofit or replacement to achieve the best conversion for your situation.

## REFERENCES

1. United Nations Environment Programme, Montreal Protocol 1991 Assessment, Refrigeration, Air Conditioning and Heat Pumps Technical Options Committee, Chapter 2, *Refrigerant Data* , UNEP, Nairobi, Kenya.

2. DuPont Technical Information Sheet, 1/94, Suva 123 (Suva Centri-LP, HCFC-123) in Chillers

3. DuPont Technical Information Sheet, 1/94, Suva 134a (Suva Cold MP, HFC-134a) in Chillers

4. United Nations Environment Programme, Montreal Protocol 1994/5 Assessment, Refrigeration, Air Conditioning and Heat Pumps Technical Options Committee, Draft Chapter 7, *Air Conditioning and Heat Pumps*, July 1994. UNEP, Nairobi, Kenya.

# Section IV

# WHETHER TO SAY GOODBYE

# CHAPTER 11

# THE MONEY GAME

You have now reached the point where you will need to make some decisions regarding your existing equipment. One of the goals of refrigerant management planning is to ensure that these decisions are made in the most cost-effective manner possible. Minimizing the upfront or first costs of the project is important. However, any business manager knows that the option with the lowest price tag may not be the option that provides the greatest value in the long term. It is important to identify projects that may save more in the long run than they cost initially. To determine both current and future costs and benefits of an action, you will need to take a long-term approach to the problem.

## TOTAL COST APPROACH

The total cost approach takes into account not only what it will cost in terms of initial outlay for retrofitting or replacing a piece of equipment, but also what additional costs or benefits may accrue over the lifetime for which those changes are in place. For example, increased energy efficiency from replacing an old chiller may provide enough savings in annual operating costs to essentially "pay off" the chiller within 4 to 5 years or less. The total cost approach provides a mechanism to evaluate the costs of various actions to choose the most cost effective approach.

The total cost approach addressed in this book is based on the following simple formula:

TCC = I + O + E

where:    I  = the investment cost
          O = the operations and maintenance costs
and       E = the energy cost.

As with any type of analysis, the accuracy of the total cost equation will depend on the accuracy and amount of data provided for each of the costs. For example, if you have accurate information on the amount of energy your existing system is using, your total cost analysis will be much more accurate than if you are using estimates based on the expected performance of the equipment. The following chapters will provide some insights into the type of information that makes up investment, operations and maintenance, and energy costs. You should keep in mind that the goal of this analysis is not to necessarily determine actual costs of implementing a particular action, but to provide a basis for an initial determination of whether an action will be cost-effective.

The equation used above assumes that all of the costs are taking place over the same time period. The operations and maintenance costs and the energy costs are the annual costs associated with these activities. The investment cost can be assumed to be a lump sum paid within a single year (the year that the retrofit or replacement takes place), or can be assumed to be a yearly lease cost. In order to accurately reflect both costs and the value of the money being spent, you need to add to the equation a calculation that takes into account the time value of money.

## THE TIME VALUE OF MONEY

In many cases, the analysis used to make an initial determination of whether an action is a good idea will assume that the value of your money today is the same as it would be if you spent it in the future. Most economists, and many business people, will acknowledge that this is not really true. Although a dollar may be a dollar may be a dollar, the one that you have in your hand right now for spending is more valuable to you than the one that you will earn next week and be able to spend. This is termed the *present value* of money, and becomes important in calculations of investments that will extend over long periods of time (like a piece of equipment with a useful life of 25-30 years), or in cases where the benefits from the investment will occur at a later date than the investment itself. In this case, the investment is the cost associated with retaining, retrofitting or replacing your existing equipment.

The calculation of present value makes the computation of the investment more complicated, but it provides the most accurate information on what your decision is really going to cost you—and what the tradeoffs are between actions (such as the difference between higher initial investment costs versus lower future maintenance costs). The formula for calculating present value is:

$$PV = \Sigma 1/(1 + d)^t] * C$$

where d  = the discount (or interest) rate used
      t  = the time period of the analysis
and   C = the cost of the investment.

In this formula, C includes all the costs that make up the total cost analysis. It includes the energy costs and operations and maintenance costs associated with your existing equipment, and the investment costs of any changes made to that equipment. The present value is calculated for each action (retain, retrofit, or replace) using the same time period, t. The example below shows how the present value and the total cost calculation go together to reflect the costs associated with your decision.

$$PV = \Sigma 1/(1+d)^t] * (\text{Investment Cost} + \text{Energy Cost} + \text{O\&M Cost})$$

The discount rate (d) in the formula above is merely the interest rate you expect to associate with your investment. Think about this as if you were to put the money in a bank, instead of using it for maintaining your equipment or undertaking a replacement. You would expect to see a return on your investment. The discount rate represents the rate of return on your investment if you were not spending it on your equipment. Basically, it is the amount you could earn on your money if you were free to invest it in any way you wish. Economists term this the "opportunity cost" of the money.*

Although the present-value calculation may look imposing, it is actually very easy to calculate, as we will demonstrate in the chapter at the end of this section using some examples of equipment retrofit and replacement decisions. But perhaps you are asking yourself "Why go through all this trouble?" The answer is quite simple. You have made an investment in the refrigeration and air conditioning equipment you currently own. Although you will eventually need to replace that equipment, you don't want to replace it before you really have to. The present-value calculation is based on the assumption that in order to maximize profits, you need to minimize costs—and the retrofit or replacement of your existing equipment can represent a substantial cost. By calculating the present value of your investment over a certain time period and comparing it to the present value of the retrofit or replacement options, you are able to determine when you need to replace your equipment while still getting the most value out of your original investment.

## YEARS TO INVESTMENT PAYBACK

Once you have made a decision to retrofit or replace your equipment, you will want to know the time it will take before your investment will start paying off. This is generally termed the "simple payback" time, and is basically the amount of time it will take the cumulative savings received from increased energy efficiency or lower operations and maintenance costs to equal the amount of the investment. The simple payback formula is:

$$SP = InvestmentCost/)Cost_{existing} - Cost_r)$$

where the investment cost is the cost of your equipment paid at a single time period, and $Cost_r$ is the cost of your replacement or retrofitted equipment. The simple payback formula assumes that the costs and savings are the same from year to year. You will notice that this simple payback calculation is not discounted. It is used only as a rough estimate of the amount of time it will take to realize a return on your investment.

## DETERMINING COSTS

All of the formulas presented in this chapter require you to know certain costs associated with both the existing equipment and the retrofitted or replaced equipment. What goes into the calculation of these costs may not be as straightforward as it seems. The following chapters will run you through some of the types of issues associated with equipment costs. Then the final chapter will help you pull all this information together to compare your options.

*There is a lot of discussion about discount rates and what rate to use. We will not go into that here. The discount rate used in this book is 7%—which is the rate that the Federal government currently directs Federal agencies to use in making their life-cycle cost evaluations. Other discount rates can be equally justified.

# CHAPTER 12

# READING THE PRICE TAG

What actually goes into determining what retaining, retrofitting, or replacing your equipment will cost in terms of capital investment? Of course, there is the cost of a new piece of equipment if you decide to replace. But there are other costs associated with your decision that also need to be considered.

## EQUIPMENT COSTS

The actual cost of the equipment being purchased or the cost of the retrofit will be included in a determination of the investment cost. For a retrofit, you will certainly need to make some changes to your system beyond simply changing the refrigerant. All of the alternative refrigerants being developed as retrofits require some changes to be made to lubricants, seals, or other components. In addition, you may also be faced with making modifications or changes to your drive motor or compressor components, such as the impeller. Depending on the retrofit, you could have minimal additions to the equipment component of your retrofit costs, or you could find that the costs are extensive.

Even if you retain your existing equipment, it is likely that you will need to, or want to, make changes to it. Unless your equipment is very new, you will probably need to invest in leak prevention. You will also potentially need to purchase recovery/recycling equipment for complying with the requirements of the Clean Air Act.

## MECHANICAL ROOM ISSUES

In addition to the costs associated with the equipment itself, there are costs associated with installing the equipment in your mechanical room. The choice of where to install the new system will have associated costs involving the actual housing of the equipment plus any additional equipment upgrades required by various applicable health and safety codes. Even if you are only retrofitting your existing equipment, you may have to examine the additional costs associated with upgrading your equipment room to meet the latest health and safety requirements that may come into effect with a new project implementation.

## SIZE

One of the things that comes to mind when examining size issues is the story of a person who builds a beautiful boat in his basement. The person does a great job and then finds that he must tear down the house to get the boat to the water. Whether you can remove your old equipment and install the new, given the constraints of your building and mechanical room, is one of the issues that can significantly raise your investment costs.

In considering a new system, the physical location of the current equipment and the anticipated location of the new equipment must be carefully reviewed before a final decision can be reached. One of the first and most obvious questions to ask is, "Will the equipment fit?" The physical size of the machine room and its location within the building or structure both have bearing on the purchase or modification of the system. In some circumstances, the inability to access equipment for removal and replacement may force you into an extensive retrofit, or to

looking for other equipment placement options (e.g., on the roof). Any new system will probably require some type of modification to the existing machine room. The extent of this modification will vary from case to case, but all associated costs of this modification need to be included in the investment cost calculation.

## ACCESS

The location of your machine room could afford easy access and easy modification; or it could be located in a sub-basement where any changes are very difficult and could require extensive, and expensive, structural modification. Access to move and position the equipment is another important consideration. The ability to use heavy lift machinery, necessary to move out old machinery as well as to install the new equipment must be carefully planned and reviewed. Whether you rent a crane or a helicopter will make a difference to your overall costs.

If equipment is not easily accessible, it may be necessary to disassemble it or use cutting torches to dismantle the old equipment to get it into pieces small enough to remove. This will probably also mean that major modifications to the space may be required to provide access for the new equipment and this could be very difficult and costly. In some cases abandoning the current machine room and building an entirely new space may be a cost effective choice.

## MACHINE ROOM CONFIGURATION
## AND REQUIREMENTS

One of the first requirements before a construction project begins is to secure a building permit or some authorization to do the project. There is usually a requirement to modify whatever facility is affected to meet the latest codes and standards. This is true of facility equipment rooms also. There may be several major or minor changes required when a retrofit or replacement is undertaken. The local codes should be examined to determine the requirements that must be met for the project to be approved. These will need to be included in the overall investment cost. Some of the most likely requirements are covered in the following discussions.

Changing your refrigerant may require you

to make changes in the ventilation of your machine room. If your machine room conforms to the requirements for construction, installation, and operation of refrigerant equipment in the Uniform Mechanical Codes and ASHRAE Standard 15-92, *Safety Code for Mechanical Refrigeration,* you will likely have very minimal changes to make after switching to one of the direct alternative refrigerants (either HFCs or HCFCs). ASHRAE modified Standard 15 in 1994 to require refrigerant sensors and monitors in mechanical rooms for all refrigerants - without regard to the type of refrigerant used.[1] The requirements for an oxygen depletion monitor has been eliminated. In addition to national standards, you will also need to be cognizant of requirements under local building codes. These codes are often more stringent than national ones.

Depending on the type of equipment being used, the fuel or energy source, and the refrigerant, the mechanical room must meet certain special safety standards. For example, ASHRAE Standard 15-1994 calls for special drains for loss of fluids, special purge and pressure-relief piping, refrigerant leak sensors and alarms, self-contained breathing apparatus (SCBA) and other health and safety equipment. If you are changing refrigerants or fuel types, you may need to upgrade some of the safety components of your machine room. This will add additional cost to your retrofit or replacement.

If the mechanical room contains both heating and cooling equipment, other considerations come into play. New rules require installing a refrigerant-sensor system that will shut down a boiler's burner when a specific concentration of refrigerant is reached in the mixed-use mechanical room. The characteristics of the alternative refrigerant, including flammability, and the fuel source must be given careful consideration in the mixed-use machinery room. Whether you decide to retain, retrofit, or replace your equipment, you may need to install burner shut down equipment if you currently have, or plan to have, a mixed-use machinery room.

The capacity of the equipment room to provide ventilation and prevent heat build up from the machinery is another factor to address when making decisions about your equipment. Many parts of the equipment, including the compressor drive and pump motors, generate heat. This heat can build up if there is insufficient ventila-

tion. If the compressor is an electric open drive, where the motor is cooled by the air in the mechanical room, the room must have sufficient air flow to control the temperature in the space to permit efficient drive-motor operation. In addition to the compressor drive motor, any coolant pump or circulating motors will also add heat to the space. Condensation occurring on the "plumbing" will also add latent heat to the space and may need control of the humidity (latent heat is also discussed in Chapter 8 on Cooling Loads). If the replacement or retrofitted equipment will increase the mechanical room heat load through any of these pathways, you may need to identify methods for increasing the ventilation of your mechanical room.

## ADDITIONAL INVESTMENT COSTS

If the system uses cooling towers, modifications made to the tower size and weight may present a significant issue in situations where the cooling tower is located on a roof or other weight limited structure. The use of "penthouses" for air handling equipment and heat exchangers may need to be reviewed, as it presents unique considerations compared to a ground-level installation. Changing the location of your cooling tower may have **considerable** associated costs.

The structural requirements or limits of the structure in which the equipment is housed may also be important in other situations. Not-in-kind replacements of refrigeration or air conditioning equipment may add to the size, weight, power supply, energy-efficiency, and installation requirements. The location of equipment and equipment components (e.g., condensing units, heat pumps, etc.) must be carefully considered in relation to the structural requirements of the building (roof supports, piping runs, access for maintenance, protection from the elements and other hazards, needed energy sources, etc.). Any costs for structural modification will have to be included in the investment cost.

Another consideration to be included in the overall investment are any expenses incurred from interrupting regular operations during the retrofit or replacement project. Will there be a need for auxiliary cooling while the installation or retrofit is taking place? If the cooling load can not be picked up by some other internal means, additional cooling capacity may need to be purchased from another source. This cost should/could be included in the investment cost. It is recommended that retrofits or replacements be timed to coincide with standard equipment service and maintenance to reduce the costs associated with equipment down-time.

## CONCLUSIONS

The considerations of all the costs associated with your decision must be carefully weighed and included in your investment cost. The textbox case study is an example of how investment costs can escalate. These calculations of the investment costs may play a significant role in determining the overall cost effectiveness of the available options.

## REFERENCE

1. 1994 Updates to ASHRAE Standard 15, Safety Code for Mechanical Refrigeration, ASHRAE, Altanta, Georgia.

---

*Scenario:** A 30-story, 400,000 square foot office building which was still running on its original HVAC equipment from 1966. The equipment was located in a "penthouse" room on the roof.

*Equipment:* The major portion of cooling was done by two 750 ton chillers. The dated equipment, which had no record of routine preventive maintenance, was obviously in need of more than just minor retrofits to keep running effectively and efficiently. To completely replace the equipment, the roof would have to be opened, the old chillers carried out by helicopters, and new machines helicoptered in. Replacement of the equipment was estimated at $1.25 million.

*The Solution:* Rather than spend over $1 million to install new equipment, the old equipment got a major overhaul. When all was said and done, they were practically new equipment running in the old shells. This included replacing the old compressors with new ones that are compatible with HCFC-123. Compressors, which are much smaller than entire units, can be easily transported up freight elevators rather than using helicopters. Next, all the evaporators and condensers got new tubing, which helped boost heat transfer efficiency and capacity. Third, purge units and chiller pressurization systems were installed. Next, new microelectronic control panels were installed. These control panels help the technicians identify and repair problems before they cause a disruption in service. The cooling tower was also retrofitted with two-speed motors. Finally, new starter panels for the compressor motors were installed to ensure smooth, dependable start-ups.

*The Results:* For only $650,000 (as compared to $1.25 million), full load energy efficiency improved from 0.90 kW/ton to 0.73 kW/ton. Partial load efficiency also increased. Further, this entire process was achieved with no disruption of service to the office building.

*The News*, June 20, 1994; Vol 192, No 8, pages 6–8. Business News Publishing Company, Troy MI

FIGURE 1.

CHAPTER $13$

# CARE AND FEEDING
## (Operations and Maintenance Costs)

The issues to be considered in determining the operations and maintenance costs for equipment also have some hidden costs. You may have fairly good information on the amount of time it takes to maintain your existing equipment - the number of inspections and repair jobs that you do over the space of a month. You may be able to get some information on expected maintenance costs from the equipment dealer to determine how much time you will need to spend with your new or retrofitted equipment. Or, you may need to make a determination about how much of the time you currently spend is based on the age or condition of your existing equipment.

New or retrofitted equipment will generally require less maintenance in some areas than your existing equipment, simply because it is newer. However, additional monitoring requirements associated with the new Clean Air Act requirements may actually increase your maintenance in other areas—such as inspections and testing. You will need to make a best estimate of whether you expect your overall maintenance costs to decrease as the result of retrofitting or replacing your equipment and by how much. Your equipment dealer should be able to provide assistance in these areas.

## REFRIGERANT AVAILABILITY

One of the questions currently facing users of CFC-refrigerants is the long-term availability of CFCs to service existing equipment once production is halted in 1995. The Air Conditioning and Refrigeration Institute (ARI) estimated that there were several million commercial air conditioning and refrigeration systems and 80,000 commercial air conditioning chillers designed for and containing CFCs[1]. CFCs have historically only accounted for 15%-20% of industrial refrigeration, with the vast majority of the remainder of the equipment using HCFC-22[2].

The production phaseout is making the supply of CFCs increasingly limited. Chiller manufacturers estimate that there will be more than 58,000 CFC-charged units still in place after January 1, 1996[3]. These units will be vying for the limited supply of recycled or recovered refrigerant. In addition, R-12 users are competing with the automotive air-conditioning sector which has the highest leakage rate of all air-conditioning users and virtually no plans to reduce that rate. The demand for CFCs to replace refrigerant lost via leaks and servicing losses is estimated at 74 million pounds in 1996—after the last CFCs have been produced[4].

Among the costs that need to be factored into your operations and maintenance cost estimates are the cost for refrigerant and the cost for unanticipated shutdown of your equipment or operations should you not be able to obtain needed CFC supplies. If you are planning to retain your existing equipment, you can probably anticipate that the costs for refrigerant will increase over the next few years. By how much is anyone's guess and depends to a large extent on the amount of refrigerant that is available. Many refrigerant suppliers have been estimating that they will have sufficient quantities of CFC refrigerants to service existing equipment in the future. A large portion of this refrigerant is expected to be recovered from equipment that is being replaced with non-CFC using equipment. Another portion of this refrigerant is com-

ing from stockpiles that these suppliers are developing prior to the production phaseout.

Depending on how much refrigerant you are going to need and over what time frame, you may be able to contract with a supplier for the amounts that you need. There is some evidence, however, that the anticipated supplies of recovered refrigerant are not being returned to these suppliers as fast as they had expected. There are indications that some users are recovering and storing their excess refrigerant for their own future uses, rather than selling this refrigerant to a supplier.

The shortage of CFCs is anticipated to be short-term, ending once sufficient numbers of equipment owners phaseout their usage and free up recovered CFCs for others. If you have sufficient supplies of your refrigerant and are in a position to survive the shortage, you may be able to retain your equipment over the remainder of its useful life. Tightening up your systems to reduce the loss from leaks may allow you to limit your refrigerant purchases and wait out the predicted shortages.

In deciding to hope for the best in purchasing refrigerant, however, you need to keep in mind the importance of the equipment to your operations, and how much you are willing to spend on an emergency retrofit or replacement should that become necessary in the future. At this writing, equipment manufactures are running about 26 weeks behind in their orders and the lead times are likely to get longer before they can meet the demand.

## COSTS FOR REPLACEMENT REFRIGERANTS

Unlike CFCs, the costs for most other alternative refrigerants is actually predicted to decrease. The prices on some HCFCs have already fallen as more equipment using these refrigerants is manufactured and put into use. At this point, it is too early to say what the final price of alternative refrigerants will be. As with CFCs, this will depend on the demand for the refrigerants, the amount of competition for manufacturing, and the general whims of the market. It is unlikely that manufactured alternatives will ever be as inexpensive as CFCs were in their heyday - but then neither will CFCs. Natural refrigerants are generally less expensive, but have more limited application than the manufactured refrigerants.

## CONCLUSIONS

The cost and availability of refrigerant is a wildcard that will surely affect decisions about retaining, retrofitting, or replacing equipment. You may have equipment with a remaining useful life of 10 years, but if you cannot afford or obtain the refrigerants that the equipment uses, you may have no option but to retrofit or replace. Added into this cost consideration is the potential cost of emergency retrofit or replacement and implications that prolonged downtime will have on operations. These issues need to be added in when calculating the O&M costs for the total cost calculation.

## REFERENCES

1. Gushee, D.E. 1993. "CFC Phaseout: Future Problem for Air Conditioning Equipment?" *CRS Report for Congress*. Washington, D.C.

2. United Nations Environment Programme, Montreal Protocol 1994/5 Assessment, Refrigeration, Air Conditioning and Heat Pumps Technical Options Committee, Draft Chapter 6, Industrial Refrigeration. July 1994. UNEP, Nairobi, Kenya

3. Davis, L. May 2, 1994. "Don't Lose Your Cool Over Refrigerant Leaks." *The Air Conditioning, Heating, and Refrigeration News*. Vol. 192. No. 1. Business News Publishing Co., Troy, Michigan.

4. Mahoney, T.A. April 11, 1994. "Statistical Panorama—Refrigerants." *The Air Conditioning, Heating, and Refrigeration News*. Vol. 191, No. 15. Business News Publishing Co., Troy, Michigan.

# CHAPTER 14

# PROVIDING THE POWER
## (Energy Efficiency)

One of the cost parameters used in evaluating decisions about what to do with your existing equipment is the amount of energy it uses, or its energy efficiency. There has been a great deal of discussion in the press and the trade literature regarding the energy impacts of the various retrofit alternatives, the new refrigerants, and the alternative technologies, such as absorption, heat pumps, and gas fired chillers. The impression left by many of the discussions is that the "new" refrigerants will result in an energy penalty because they are less efficient. This issue has several facets that need to be explored.

The vapor-compression cycle takes advantage of the thermophysical properties of a refrigerant to achieve a transfer of heat energy. These properties, to be useful, must satisfy certain thermodynamic and transport requirements. Generally speaking, there will not be a single refrigerant that covers the temperature range, capacity, and efficiency of all refrigeration and air conditioning requirements. Rather, the choice of a refrigerant balances tradeoffs in properties to satisfy a given application. One of the most critical issues involves balancing the efficiency of a cycle and its cooling capacity. As a rule, the efficiency of a refrigerant increases with the increasing critical temperature while the capacity decreases[1]. The efficiency of a refrigerant influences the energy requirements directly, while the capacity affects the actual capital cost of the equipment.

The current halocarbon refrigerants have been used for quite a long time, and over the years, refrigeration and air conditioning equipment has been optimized for use with these refrigerants. With the advent of the phaseout, there has been extensive work on trying to develop a "drop in" replacement for CFCs and HCFCs that would require no changes to the current systems, and work as well as the refrigerant being replaced.

Some replacements have been developed to be used as retrofits for existing equipment with minimal changes to the system. However, the use of these new refrigerants in a system optimized for use with halocarbons has aroused concerns about the energy efficiency of such retrofits. For example, in 1990 when a machine was converted from CFC-11 to R-123, the owner could expect a 20% loss in cooling capacity and a 10% loss in efficiency.[2] Today, this is not necessarily the case. System engineers have discovered that they can reduce the losses in efficiency and capacity resulting from the different thermodynamic properties of the replacement refrigerant by optimizing other components of the system, such as the compressor, the drive train and motor, the condenser, and the evaporators. This can reduce the expected capacity losses to between 0% and 5%, and efficiency losses to between 0% and 4% of the original system. While this increases the number of components that need to be changed in a retrofit, it can ensure that capacity and energy efficiency losses are kept to a minimum.

The phaseout of the halocarbons and the development of replacement equipment that utilizes alternative refrigerants have provided the opportunity to design replacement equipment that is optimized for these new refrigerants. The new machines that are being developed are more efficient than the machines they replace. A recent study by Trane has indicated that the overall efficiency of some systems has gone from .86 kW/ton in 1970 to .52 kW/ton in 1994 (Figure 1).

## COEFFICIENT OF PERFORMANCE AND EFFICIENCY MEASUREMENTS

The "energy efficiency" of a piece of equipment is a function of many variables. Different types of equipment and different refrigerants will have different levels of performance. On a basic level, a refrigeration cycle can be thought of as a continuous transfer of thermal energy (heat) from a low temperature area to a higher-temperature area. The low temperature area is the space to be cooled, and the higher-temperature area, or "heat sink," is usually cooling water or ambient air. The efficiency of a piece of equipment is defined as the energy output (amount of cooling) divided by the energy input (electricity or other energy consumption) under full-load conditions of the equipment. The energy performance of refrigeration and air conditioning equipment is usually expressed as the coefficient of performance (COP). The COP is a ratio of the rate of heat removal and the rate of energy use:

COP = rate heat removal / rate of energy input

COPs for equipment are usually specified by the equipment manufacturer or designer. A sample of some ranges of COPs for equipment is shown in Figure 2. Another common way to compare efficiencies or performance of equip-

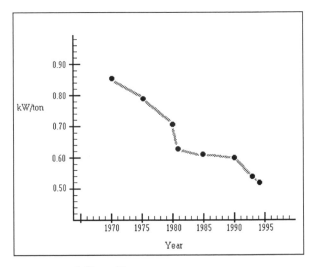

*FIGURE 1. Chiller Efficiency.* (Courtesy Trane, Inc.)

| Equipment Type | High COP | LOW COP |
|---|---|---|
| Absorption | 1.8 | 0.4 |
| Contrifugal | 6.75 | 2.5 |
| Reciprocating (electric) | 3.5 | 2 |
| Reciprocating (gas) | 1.9 | 1.5 |
| Screw Compressor | 5 | 3 |

FIGURE 2. Ranges of coefficient of performance (COP)* for various technologies.(advertised or repeated from various sources)

ment is by using kilowatts per ton (kW/t). This is basically the amount of cooling produced by the evaporator, divided by the input energy required by the compressor. The kW/t of a piece of equipment can be calculated from the COP.

Performance (kW/t) = 3.516/COP

For Example the performance of a machine that is rated with a COP of 4.5:

kW/ton = 3.516/4.5

= .78 kW/ton

This book uses, kW/t to compare the energy efficiency of various equipment options.

## ENERGY COSTS

The energy cost for operating your equipment is the single largest cost component in your analysis. Therefore, on a purely economic basis, a retrofit or replacement option that increases your energy costs will end up increasing your overall costs. Even a small change in your energy efficiency can provide significant changes in your annual energy costs. For example, assume you have a 700 ton chiller with an efficiency of .79 kW/t. It operates an estimated 4000 hours per year, and you know that the cost for electricity is .08$/kWh. You can calculate your current energy usage in the following manner.

700 tons * .79 kW/t * 4000 hrs/yr * 0.08$/hr = $176,960.00

Now assume that you make a change to your system that results in an energy efficiency gain that gives you an efficiency of .75 kW/t. Using the same calculation you get an annual energy cost of $168,000.00. This is an annual energy savings of $8,960.00. If you assume that the energy efficiency goes down rather than up—for example ,your efficiency is .81 kW/t—your annual energy cost goes up to $181,440.00. This is an annual increase in energy cost of $4,480.00.

In some situations the reduction in energy expenditures can actually save enough to allow a facility to "pay back" the cost of a new system or a retrofit very quickly in relative terms. In many cases where the equipment is old and the efficiency is low, a replacement could achieve a payback in a very short time. In addition, there may be additional energy gains that can be realized by switching from one fuel to another. Changing from electric power to other fuel sources can be a complicated endeavor. The simplest change is one from electric power to natural gas or vice versa. The motivation for this change could be based on utility incentives or cost analysis of utility rates and charges. Other fuels that can be and have been used to power refrigeration and air conditioning equipment include geothermal, waste steam/heat, solar, and petroleum. In fact, many industrial absorption systems are powered on waste energy—making these systems an attractive option for industries that create waste heat as a by-product of their activities.

## DEMAND-SIDE MANAGEMENT INCENTIVES

One of the more interesting issues in the energy management area is known as Demand-Side Management (DSM). This is a program where the local utility (usually electric) provides various incentives to its customers to encourage them to install more energy efficient equipment or to take other steps to conserve energy. The primary reason for this program is to delay or eliminate the need for building additional electric-generating capacity. The capital costs of new generating plants, as well as the environmental impacts and issues involved, make the building of new generating facilities enormously expensive.

While many of the special DSM programs are aimed at the residential sector, there are programs available to the commercial and industrial sectors. The incentives can be implemented in a number of ways. Some of the novel incentives include utility-provided maintenance services for HVAC systems including comprehensive large-system diagnostic, maintenance, and training services. Installing large-scale building-wide monitoring and control systems for large commercial buildings is another service utilities promote through incentives. One utility has what is referred to as the "Chiller Early Retirement Program," where incentives are supplied for chiller replacement, but this is also tied to a requirement for the customer to commit to a high-efficiency lighting retrofit program.

Most of the incentives involve rebate payments to the customers in the form of $/kW reduced. Other approaches include covering a percentage of the incremental cost of equipment improvement or upgrades. These can range from 11% up to 100% of the costs. Some of the utility incentives are tailored to individual customers to produce one and two-year paybacks. In addition to the direct rebates, some utilities offer special low-interest or third party financing as a single incentive or in conjunction with other rebate payments.

It is obvious that before any project is begun, a check with the local utility could be very profitable. For example, a recent effort in the Wisconsin Electric Power Company service area started out as a $200/kW-offset and netted the customer $890,000 in rebates involving a system supplying 1400 tons of cooling capacity[3]. This simple example shows that if the utility does not have a DSM program, then it might be time to convince them to start one.

## CONCLUSIONS

This chapter has identified some of the costs and benefits that should be included in the calculation of the energy cost for evaluating retain, retrofit, and replace options. Although energy is the largest single annual cost associated with your equipment, decisions on whether to retrofit or replace will need to take into account the other factors mentioned in this book. The ability to realize energy efficiency gains, however, provides a significant incentive for making changes to your equipment. Additional cost savings from fuel switching or demand-side management programs should also be factored into your energy cost analysis.

## REFERENCES

1. United Nations Environment Programme, Montreal Protocol 1994/5 Assessment, Refrigeration, Air Conditioning and Heat Pumps Technical Options Committee, Draft Chapter 2, *Refrigerant Data,* July 1994. UNEP, Nairobi, Kenya.

2. E.L. Smithheart, "Choosing a Building Chiller." Proceedings, *International CFC Conference and Halon Alternatives Conference,* October, 1993, Washington, D.C. Sponsored by the Alliance for A Responsible CFC Policy, Frederick, MD.

3. Engineered Systems, *Thermal Storage By Design,* July 94, Vol 11, No 7, Business News Publishing Co. Troy, Michigan.

# CHAPTER 15

# PUTTING IT ALL TOGETHER

Now that you have gone through this book and read about the issues to consider in dealing with your ozone-depleting refrigerants, you may be asking yourself, "How do I pull this all together?" How **do** you make a decision whether to retain, retrofit, or replace your equipment? How do you determine the best time to make that change?

The goal of this chapter is to provide some examples of how the information provided in this book can be used to make informed decisions about your equipment. As with the other chapters, it is designed to provide an estimate of "best guess," given information on your existing equipment and on the retrofit or replacement options you are considering.

## START WITH WHAT YOU HAVE

At the beginning of section II you started an inventory of information on your existing equipment. We are going to use some of the informa-

91

tion in that inventory now to determine what you should do with your equipment. The parts of the inventory that we don't use in this decision-making section are still useful. Information on the manufacturer and model number, for example, will be important in determining how to move once a decision is made.

The inventory information from a hypothetical chiller is provided in Figure 1. This is the piece of existing equipment we will be examining in our examples. You can use the forms at the end of this chapter to replace our example information with your own real data and follow along with us.

## RETROFIT?

The first question to ask yourself when deciding whether to retrofit your equipment is whether there is a retrofit available. For our example of an R-11 chiller, the only retrofit option is R-123. If the chiller was an R-12 or R-502 type, you may be able to use one of the blends discussed in Chapter 9. Go back to the information you recorded at the end of Chapter 9 and determine if retrofitting is an option for you.

After you have determined that a retrofit option **is** available, the second question to ask is what is the remaining useful life of your equipment. Equipment manufacturers generally use 25 to 30 years as the lifetime of a piece of refrigeration or air conditioning equipment. There are probably a lot of pieces of equipment out there that exceed this time frame, and this is OK

as long as the equipment is in good operating condition. Because most retrofits that will provide energy-efficiency savings involve replacement of some of the major moving parts (e.g., the motor), a retrofit can be expected to add a few years to the remaining useful life of your equipment. At most, you can probably expect the retrofit to increase your equipment's life span by 10 years—although 5 years is probably more realistic for most circumstances.

The ability to achieve energy efficiency gains will be greater the older your equipment is. Figure 2 shows the energy efficiency gains that have been realized in equipment over the last several years. On the other hand, the older your equipment, generally, the poorer its condition will be. In general, retrofits will be most optimal for equipment that is relatively new, out to equipment that has about 10 years remaining useful life. This is just a ballpark figure, however, and you will need to judge your own equipment on its merits.

Another issue of great import in determining whether to retrofit or replace your equipment is the equipment's accessibility. As the example in Chapter 12 indicated, the need to make major modifications to the building can quickly price a replacement out of the market. Most retrofits can be installed using existing service corridors, and some components may be able to be disassembled to make conversion easier. If your equipment is very inaccessible, a retrofit may be your only option.

For our example, we have chosen a piece of equipment that is 15 years old. By subtracting 15 from a useful life of 25 years, we have a re-

**EXISTING EQUIPMENT INFORMATION**

| | |
|---|---|
| Type of Equipment | Centrigugal Chiller |
| Age of Equipment | 15 years |
| Type of Refrigerant | R-11 |
| Charge | 2500 pounds |
| COP or kW/t | 0.79 kW/t |
| Electricity Cost ($/kWh) | .08 $/kWh |
| Gas Cost ($/Therm) | .40 $/Therm |
| Estimated Annual Hours of Operation | 6,000 |
| Estimated Leakage Rate | 10%/year |

*FIGURE 1.*

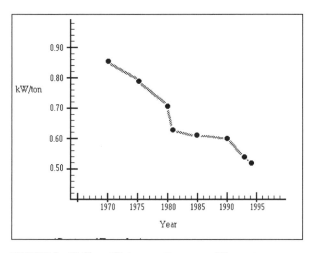

*FIGURE 2. Chiller efficiency.* (Courtesy of Trane, Inc.)

## INVESTMENT COSTS

What goes into the investment cost for our hypothetical retrofit? This can be broken down into several parts:

### Equipment

R-123 has similar characteristics to R-11, but it is a stronger solvent. We plan to replace the existing gaskets, O-rings and seals with parts that are more compatible with R-123. We anticipate that the lubricant we are currently using for R-11 will also be compatible with R-123. We will use the opportunity of the retrofit to install high-efficiency purge units on our system. These units will significantly reduce refrigerant emissions. We also plan to install a pressurizing system that will create a pressure seal on the equipment to prevent air, moisture and other non-condensibles from leaking into the system when it is shut down. At the same time, we will install isolation valves on the oil filters to reduce refrigerant loss during oil changes.

We anticipate a cooling capacity loss from the conversion of 10%. This is within the 5% to 20% range provided by the refrigerant manufacturer. From our estimate in Chapter 8, we have determined that our capacity exceeds our cur-

maining useful life of 10 years. This means, all things being equal, we could expect our equipment to last another 10 years before needing replacement.

rent cooling load requirements by 15%, so this additional capacity loss should not have any anticipated effect on our operations. If we were currently at our required cooling capacity, the 10% loss might require us to make changes to the impellers or other components to bring the capacity of the system up to our needs.

The refrigerant manufacturer indicates that a straight retrofit of the equipment, without optimizing the impeller and refrigerant flow orifices, will result in a 0% to 5% energy efficiency loss. Since this would increase our annual energy costs, it does not make sense for us to undertake such a retrofit. Therefore, we will be investing in these changes to optimize the equipment to use R-123. Once accomplished, this should result in an energy efficiency saving of 2.9% to 4.4%.[1]

The total cost of the changes to the equipment are estimated to be: **$160,000.00**. For the purposes of our analysis, we will assume that changes to the equipment will be purchased on a lease that we have worked out with our vendor. We are assuming a 10 year period for this lease, although different periods could also be used. Thus, the cost will occur in annual payments of: **$24,385.23***

### Mechanical Room Changes and Ventilation Upgrades

To comply with ASHRAE Standard 15-94, we will need to install a refrigerant vapor detector in the mechanical room with an alarm system located outside the mechanical room. We will also need to install ventilation piping from the purge units to the outside air, and some ventilation exhaust fans. The cost for these changes is **$20,000.00**

### Refrigerant

Our retrofit requires us to purchase 2500 pounds of R-123. We estimate the current cost for refrigerant to be **$6.00/lb**[2]. However,

---

*The company from which you will purchase your equipment will determine the yearly or monthly payments for you. For the examples used here, we assumed a loan at 8.5% payable over the next 10 years. We then used the following formula:

$$PMT = INV * i/1 - (1+i)^{-N}$$

because we are recovering the R-11 that we are removing from the system, we plan to receive a credit for returning this R-11 to our refrigerant supplier. At present, the resale cost for R-11 is **$10.00/lb[3]**. The total refrigerant cost is expected to be **$15,000.00**. The anticipated credit for our R-11 is **$25,000.00**.

## OPERATIONS AND MAINTENANCE COSTS

We expect increased maintenance as a result of the requirements for leak reduction and recycling/recovery activity in the Clean Air Act. There are also additional record-keeping requirements. However, as a result of the high-efficiency purges and new seals, gaskets, and O-rings, we anticipate operational costs to be reduced somewhat. The increased efficiency of the purge units—which will reduce our refrigerant losses to 1% or less—make the cost of replacement refrigerant negligible. The overall annual operations and maintenance costs for our converted chiller are estimated to be **$18,750.00/yr.**

### Energy

From the inventory information that we collected on our existing equipment, we know the rated energy efficiency in kilowatts per ton (kW/t)—.79 kW/t. A simple conversion from R-11 to R-123 is expected to result in a 0% to 5% loss in COP over R-11. However, we have undertaken an engineered conversion to prevent this loss. The analysis performed on our optimized conversion by the contractor indicates we will actually increase our energy efficiency by 2.9%. The text box below illustrates how to incorporate the 2.9% efficiency gain.

---

**A Formula for Calculating Efficiency Change**

$$= \frac{1}{1 + \% \text{ efficiency}} * \text{current kW/t}$$

Efficiency Change

$$\frac{1}{1 + 0.029}$$
$$= 0.972$$

Retrofit kW/t

$$* \frac{\begin{array}{c} 0.79 \\ 0.972 \end{array}}{0.768 \text{ kW/t}}$$

---

If we only had the COP for our existing equipment, we could use the equation from Chapter 14 to convert the COP to kilowatts per ton (kW/t):

$$kWt = 3.516/COP = 3.516/4.45 = .79$$

The kilowatt hour energy cost in our area is .08$/kWh. We also know that our equipment operates on average 6,000 hours annually. This information was collected when we performed the inventory. To determine the annual energy consumption of our retrofit, we can use the following formula:

$$Size(tons)*kW/t*annual operating hours*\$/kWh$$

Using our information, this results as below:

$$100 tons*.767 kW/t*6000 hrs/yr*0.08\$kWh$$

Thus we see that the total annual energy cost for our example is **$368,160.00**.

## COST ESTIMATE

We now have all the information that we need to determine the cost of our retrofitted equipment over the next ten years of its life. This is shown in Figure 3. We reintroduce the formula provided in Chapter 11.

$$PV = \Sigma[1/(1 + d)^t]*C$$

We will use this formula to calculate the overall costs for each of the ten years, and then sum these costs together to get the present value calculation of the total costs for the project during that 10-year time frame. If you have a calculator that will perform summation, you can simplify this calculation. The investment costs for the first year of the calculation include the first year of our lease costs, plus the total cost of the changes made to the mechanical room, plus the cost of the refrigerant minus the resale value. The investment cost for the following years will be only the annual lease cost. We will use a 10-year time frame so that we can make a comparison with the option to retain the equipment over the remainder of its useful life. We have shown the steps in the calculation in Figure 4.

## CONCLUSION

You now know how much it will cost over

FIGURE 3.

*RETROFIT EXAMPLE:*

| | |
|---|---:|
| *Investment Cost: (first year)* | *$33,385.23* |
| *Investment Cost (following years)* | *$24,385.23* |
| Equipment | $24,385.23 |
|    O-rings, seals gaskets | |
|    High efficiency purges | |
|    Pressurizing System | |
| Mechanical Room/Ventilation (single year) | $20,000.00 |
|    Vapor Detector | |
|    PurgePiping | |
| Refrigerant (initial charge) | $14,000.00 |
| R-11 Credit (single year) | ($25,000.00) |
| *Operations and Maintenance Cost: (yearly)* | *$18,750.00* |
| *Energy Cost: (yearly)* | *$368,160.00* |
|    COP    4.58 | |
|    Kw/t   0.767 | |
| *Total Cost First Year* | *$420,295.23* |
| *Total Annual Cost Succeeding Years* | *$411,295.23* |

FIGURE 4.

*RETROFIT EXAMPLE: Cost Calculation*

| | |
|---|---:|
| *Investment Cost: (first year)* | *$33,385.23* |
| *Investment Cost: (following years)* | *$24,385.23* |
| *Operations and Maintenance Cost: (yearly)* | *$18,750.00* |
| *Energy Cost: (yearly)* | *$368,160.00* |
| *Total Cost First Year* | *$420,295.23* |
| *Total Annual Cost Succeeding Years* | *$411,295.23* |

| | |
|---|---|
| Year 1: | $392,799.28 |
| Year 2: | $384,388.07 |
| Year 3: | $335,739.42 |
| Year 4: | $313,775.16 |
| Year 5: | $293,247.81 |
| Year 6: | $274,063.38 |
| Year 7: | $256,134.00 |
| Year 8: | $239,377.57 |
| Year 9: | $223,717.35 |
| Year 10: | $209,081.64 |

| | |
|---|---|
| *Total Cost Over 10 Years* | *$2,922,323.68* |

the next ten years to retrofit your equipment. In order to know whether retrofitting is more cost-effective than replacing or retaining the equipment, you need to perform calculations for those options as well. You will then be able to make a comparison between them.

## REPLACE?

As discussed previously, one of the primary considerations in deciding whether to replace your existing equipment is its accessibility. For some particularly inaccessible equipment, it is possible you may not be able to move the old equipment out and a new piece of equipment in. At this point, you may end up undertaking so significant a conversion that it will, in the end result, be a replacement of the existing equipment. For the purposes of the calculations, however, we will address equipment replacement as the physical replacement of one system with another complete system.

The replacement example we will use will be to replace our existing R-11 chiller with one that uses R-134a. We could easily use R-123 for our replacement as well as our retrofit, but we chose to use R-134a for illustration purposes. We could also replace our existing equipment with a not-in-kind technology, such as absorption. The type of system that you chose for replacement will be based on your specific application and the availability and efficiency of alternatives for that equipment.

## INVESTMENT COSTS

As with the retrofit example, there are several components that go into the cost factors for our replacement.

## Equipment

Because this is a new piece of equipment, the only equipment-related costs that we will have is the actual cost of the equipment, and the cost for installation. The estimated cost of a new R-134a chiller in our size range is: **$294,400.00**. As with the retrofit, we are assuming that the cost for the equipment and installation is paid via a lease agreement over the next 10 years. The annual payment turns out to be: **$44,868.83**.

The temperature range of our application is such that we do not anticipate any capacity

losses with our new R-134a chiller; the new chiller is optimized for use with R-134a, in contrast to the retrofit situation where the chiller was optimized to use R-11. We also expect the chiller to be physically about 30% smaller than our existing chiller based on manufacturer indications.

## Mechanical Room and Ventilation

We still need to modify our mechanical room to comply with ASHRAE 15-94, which provides the same requirements for all refrigerants. Thus, the anticipated costs for mechanical room changes are the same for those in the retrofit example: **$20,000.00**.

## Refrigerant

We will need to purchase 2,500 pounds of R-134a for our new chiller. As with the retrofit option, we should also receive a credit on the R-11 that we are able to sell back to the distributor. Using a current rate of **$7.00/lb**[4] for R-134a, the anticipated cost for refrigerant is **$17,500.00**. The anticipated credit from sale of our recovered R-11 is **$25,000.00**.

## OPERATIONS AND MAINTENANCE COSTS

Because this will be a new chiller, we may expect to see a decrease in operations and maintenance costs. As with the retrofit example above, we will have some increase in operations and maintenance over our existing chiller if we have not previously recovered and recycled our refrigerant during servicing. We will also have increased recordkeeping. However, because the system is new, we can anticipate a decrease in general maintenance. Based on an estimated 57 hours per month in maintenance, we estimate our operations and maintenance costs to be **$17,100.00**.

## ENERGY

The manufacturer of our new R-134a chiller indicates that it will have a COP of 5.7. We can use the same analysis applied to the retrofit to convert the COP to kW/t:

$$kW/t = 3.516/COP = 3.516/5.7 = .617$$

and then calculate our energy cost:

$$1000 tons * .62 kW/t * 6000 hrs/yr * 0.08 \$/kWh$$

Based on this calculation, we anticipate our annual energy costs to be **$297,600.00**.

## COST ESTIMATE:

We now a have all the information that we will need to calculate the cost estimate for a chiller replacement. This is shown in Figure 5. The anticipated useful life of the chiller is between 25 and 30 years; however, we will only calculate the first ten years of the chiller life in order to be able to make a comparison with the retrofit and replacement options. In general, the longer the period of time over which the costs are spread, the less "expensive" the replacement or retrofit will be. In order to determine the tradeoffs between options, however, we need to compare the costs over the same period of time.

As we did under the retrofit example, we will calculate the annual costs for our hypothetical chiller replacement for each of the next ten years. Then we will sum the yearly totals together to get the overall cost for the chiller during that time period. This calculation is shown in Figure 6. Again, the C parameter of the equation is a sum of all the yearly costs. The first year of the investment cost is assumed to include the first year of our lease payment, plus the changes to the mechanical room, plus the cost of the refrigerant, minus the sale of the surplus refrigerant. The investment cost for the following years will only include our lease cost.

## CONCLUSION

At this point we can see some difference in the costs over the next ten years for the retrofit and replacement options. The replacement has a slightly lower cost over the period. Although the replacement has an overall lower investment cost, the energy efficiency reduction between the replacement chiller and our old chiller is significant. This results in lower annual costs for energy, and therefore a lower overall cost for the replacement over the ten-year time frame. Another factor bearing on our analysis is the time frame for the analysis and the lease period. If we were to increase the time frame for the analysis, the replacement option would continue to become increasingly cost effective because, after then end of the 10 years, there

would be no more investment cost. Similarly, if we had chosen a shorter lease period for either of the options, we would have increased the annual costs, and thereby increased the total cost over the period.

Now let's run the calculations for retaining our equipment over the same time frame and see how it compares to the retrofit and replacement options.

## RETAIN?

Another option for our relatively young chiller is to retain it for the remainder of its useful life. The ability to do this depends in large part on the availability of refrigerant to service the equipment. During the equipment inventory, we estimated the amount of leakage that our equipment currently has. For our example, we have found a 10% leakage rate. Since the charge for our equipment is 2,500 lbs, this amounts to a loss of 250 lbs of refrigerant per year to leaks. At the current cost for R-11 of $20/lbs, this means we are spending $5,000 per year in refrigerant additions.

For this example, we are going to assume that refrigerant will be available. We may be able to do this because we have banked some refrigerant, or because we have a contract with a company that will guarantee us a refrigerant supply. In our analysis, we will assume that we will invest in some technologies to bring our leakage under control. You could also run this analysis for a shorter period of time—if, for example, you thought you might be able to obtain refrigerant for only 5 years instead of 10. In that case, to make a comparison between the retrofit and replacement options you would need to ensure that the time period used was the same for all scenarios.

## INVESTMENT COSTS

Although we are not purchasing a new piece of equipment or undertaking a retrofit, we will still need to make some changes to our current operations that will need to be counted as investment costs.

### Equipment

Given the predicted shortages in halocarbon refrigerant supplies, it is in our interest to invest

*REPLACEMENT EXAMPLE:*

| | |
|---|---|
| *Investment Cost: (first year)* | *$57,368.83* |
| *Investment Cost (following years)* | *$44,868.83* |
| Equipment | $44,868.83 |
| Mechanical Room/Ventilation | $20,000.00 |
| Refrigerant (initial charge) | $17,500.00 |
| R-11 Credit | ($25,000.00) |
| *Operations and Maintenance Cost: (yearly)* | *$17,100.00* |
| *Energy Cost: (yearly)* | *$297,600.00* |
| COP 5.67 | |
| kW/t 0.62 | |
| *Total Cost First Year* | *$372,068.83* |
| *Total Cost Succeeding Years* | *$359,568.83* |

*FIGURE 5.*

*REPLACEMENT EXAMPLE:* *Cost Calculation*

| | |
|---|---|
| *Investment Cost: (first year)* | *$57,368.83* |
| *Investment Cost (following years)* | *$44,868.53* |
| *Operations and Maintenance Cost: (yearly)* | *$17,100.00* |
| *Energy Cost: (yearly)* | *$297,600.00* |
| *Total Cost First Year* | *$372,068.83* |
| *Total Annual Cost Succeeding Years* | *$359,568.53* |

| | |
|---|---|
| Year 1: | $347,727.88 |
| Year 2: | $336,045.36 |
| Year 3: | $293,515.03 |
| Year 4: | $274,313.11 |
| Year 5: | $256,367.39 |
| Year 6: | $239,595.69 |
| Year 7: | $223,921.21 |
| Year 8: | $209,272.16 |
| Year 9: | $195,581.46 |
| Year 10: | $182,786.41 |

| | |
|---|---|
| *Total Cost Over 10 Years* | *$2,559,125.69* |

*FIGURE 6.*

in refrigerant containment equipment. Although our 10% leakage rate is well below the minimum rate for regulation under the Clean Air Act, we have also shown that we are losing $5,000 per year in refrigerant leaks. As the price for CFCs increases, we can expect our monetary loses to increase as well.

For these reasons, we are planning to install a high-efficiency purge on our equipment. We anticipate that this purge will decrease purge-related emissions to 1% or less. An inspection of the system also shows that we could stand to replace some gaskets and O-rings that are showing signs of wear. We will make plans to replace these components during the next scheduled maintenance. We will also install a pressurizing system to prevent air, moisture, and other non-condensibles from leaking into the system when it is shut down. At the same time, we will install isolation valves on the oil filters to reduce refrigerant loss during oil changes.

The anticipated cost for these changes to our hypothetical system is **$24,000.00**.

## Mechanical Room and Ventilation

Because we are not undertaking major changes to the equipment, we are not required to bring our mechanical room up to the requirements of ASHRAE 15-94. Since the option to retain our equipment includes an understanding that we will be replacing it at the end of the 10-year remaining useful life period, we have opted to wait until we identify a replacement to make the equipment room changes. Thus, we do not anticipate any costs related to mechanical room or ventilation changes.

## Refrigerant

We will not need to purchase any refrigerant at this time. However, we may wish to do so in order to "bank" refrigerants in the future. If we have reduced our leakage to 1% by addition of the containment devices discussed above, our anticipated annual consumption of R-11 is only 25 pounds. At current prices, this equates to $500 per year for refrigerant. We could bank the 250 pounds of refrigerant that we anticipate we will need over the next ten years to make up for leakage losses for only $5,000—or basically what it costs us annually for refrigerant before investing in containment. If we believe that refrigerant

costs will increase dramatically over that ten-year period, we may want to purchase that refrigerant now. Prior to doing that, however, we will need to check on state and local building codes to determine if any exist that would regulate the amount of refrigerant that we are able to store.

For our example, we are not planning to purchase this refrigerant. Even if refrigerant prices were to increase 100% over the next ten years, the overall cost for refrigerant is significantly smaller than either the investment cost or the energy costs of equipment operation as to become almost negligible in the calculations.

## OPERATIONS AND MAINTENANCE

Since our equipment is 15 years old, we are going to assume that it requires more maintenance than either the retrofitted or new equipment would. In addition, this is expected to increase based on the increased regulations under the Clean Air Act. For our hypothetical example, we are assuming that our equipment will take approximately 80 hours/month to maintain. At a labor rate of $25/hr our operations and maintenance costs are anticipated to be **$24,000.00 per year**.

## ENERGY

Based on the information provided with our existing equipment, we know that the rated efficiency is .79 kW/t. In order to determine the energy cost we can use the same formula that we used in the retrofit and replacement examples:

$$1000 tons * .79 kW/t * 6000 hrs/yr * 0.08 \$/kWh$$

Using this information, we calculate our existing annual energy costs to be **$379,200.00**.

## COST ESTIMATE

The cost estimate for our retain example is shown in Figure 7. We calculate the cost estimate over the ten years of remaining useful life in the same manner that we did for the retrofit and replacement examples. In our retain example, however, we do not have an investment cost for years 2 through 9 as we did in the retrofit and replacement examples. The total cost for changes to our equipment will be assumed to

*RETAIN EXAMPLE:*

| | |
|---|---|
| ***Investment Cost: (single year)*** | ***$24,000.00*** |
| Equipment | $24,000.00 |
|    O-rings, seals gaskets | |
|    High efficiency purges | |
|    Pressurizing System | |
| Mechanical Room/Ventilation | $0.00 |
|    Vapor Detector | |
|    Purge Piping | |
| Refrigerant (initial charge) | $0.00 |
| ***Operations and Maintenance Cost: (yearly)*** | ***$24,000.00*** |
| ***Energy Cost: (yearly)*** | ***$379,200.00*** |
|    COP    4.45 | |
|    kW/t    0.79 | |
| ***Total Cost First Year*** | ***$427,200.00*** |
| ***Total Cost Succeeding Years*** | ***$403,200.00*** |

FIGURE 7.

*RETAIN EXAMPLE: Cost Calculation*

| | |
|---|---|
| ***Investment Cost: (single year)*** | ***$24,000.00*** |
| ***Operations and Maintenance Cost: (yearly)*** | ***$24,000.00*** |
| ***Energy Cost: (yearly)*** | ***$379,200.00*** |
| ***Total Cost First Year*** | ***$427,200.00*** |
| ***Total Annual Cost Succeeding Years*** | ***$403,200.00*** |
| ***Total Cost Last Year*** | ***$372,068.83*** |

| | |
|---|---|
| Year 1: | $399,252.34 |
| Year 2: | $376,822.43 |
| Year 3: | $329,131.30 |
| Year 4: | $307,599.35 |
| Year 5: | $287,476.03 |
| Year 6: | $268,669.18 |
| Year 7: | $251,092.70 |
| Year 8: | $234,666.07 |
| Year 9: | $219,314.09 |
| Year 10: | $189,140.93 |

| | |
|---|---|
| ***Total Cost Over 10 Years*** | ***$2,863,164.41*** |

FIGURE 8.

take place in the first year. In addition, at the end of the 10-year period, our useful life will be over—and we will need to replace the equipment. Therefore, in the 10th year, we are assuming that we purchase a new piece of equipment for the same price as we have in our replacement example. The 10th-year costs then include the first year of our lease payment for the equipment. Since we did not make changes to our equipment room now, we will need to make those when the equipment is replaced. Thus, the cost in the 10th year includes the cost of those changes as well. In addition, we are going to assume for our example that the investment in the equipment is made at the beginning of the 10th year. This means that we can use the energy and operations and maintenance costs associated with the new piece of equipment. Since these costs are lower, we will want to ensure that we calculate the benefits of equipment replacement as well as the cost. The cost analysis is shown in Figure 8.

## PULL IT ALL TOGETHER

You can see by comparing the analyses for the retrofit, replace and retain examples that the costs are fairly close. The replacement option is somewhat more cost-effective over the study period, because of the significant energy efficiency gains that we obtain from the new equipment. This significantly reduces our annual costs. In general, the higher the annual cost savings that you can obtain, the more cost-effective an investment will be, even if it has a relatively high initial cost.

The retain example comes out slightly better than the retrofit option, because we have been able to defer our costs. However, this assumes that we will actually be able to obtain the needed refrigerant supplies to keep this equipment this long. In our retain example, we have also not calculated costs for purchasing refrigerant in the future. Our assumption was that the anticipated increases in CFC costs would still be small compared with the energy costs of operating the equipment, and therefore would not significantly affect our analysis. There is a point, however, at which refrigerant costs may become a factor.

In examining these examples, it is important to keep in mind some of the assumptions that have been made. In general, a retrofit tends to

be undertaken to match the requirements of the existing system. Therefore, it generally tends to match capacity or kW/t parameters. If you have read Chapter 8, you know that the capacity requirements for your equipment may have changed over time because of changes made to your building. The installation of energy efficient lights, for example, will reduce your cooling load requirements. This can be factored into the retrofit analysis to identify whether decreasing the capacity is feasible. These same issues should be kept in mind when identifying replacement equipment as well.

Another area where changes can be made that will affect your retrofit or replacement is in the temperature of the cooling water coming into the system. If you are able to decrease the cooling water temperature, this can significantly boost your cooling capacity. Of course, this change needs to be within the allowable limits for your equipment. It is important that you share information on all aspects of your system and any changes that you may anticipate with the equipment manufacturers prior to determining the type of retrofit. This will ensure that you get the optimal retrofit or replacement for your situation.

In the examples above, we used a lease option for making our major investment. This is not required. The investment cost could be paid as a lump sum in the first year of the action—whether retrofit or purchase. We encourage you to play around with these equations using different scenarios. Forms for calculating your own options are provided at the end of this chapter.

## SIMPLE PAYBACK

One of the other issues of importance in determining the benefits of an investment is the payback period. Using the formula provided in Chapter 11, we have calculated the investment payback period for the retrofit and replacement options as follows:

$$SP = Total\ Investment\ Cost/(Cost_{existing}-Cost_r)$$

The costs used in this formula are the annual costs for energy and operations and maintenance for both your existing equipment ($Cost_{existing}$) and your retrofitted or replaced equipment ($Cost_r$). The investment cost is used as a lump sum. This is different than our examples above, where we have chosen to lease our

101

equipment. The simple payback periods for the two options are shown in Figure 9. You can see that the cost savings from energy-efficiency gains for the replacement result in a payback in 3 years, while the retrofit requires 10. Again, this is based on the parameters that we specified in determining our retrofit and replacement, it may not be based on the payback you could receive with an "optimal" retrofit.

## CONCLUSIONS

This chapter has provided some examples for pulling together the information on your various options to make some financial comparisons between them. There are many variables that can be added to each of the scenarios used above to make them applicable to your individual situation. In addition, the examples did not address not-in-kind replacements, which also might be valid options for some circumstances.

It is important to keep in mind that the assumptions made when sketching out your scenarios are going to influence the results of the analysis. There are no good options or bad

options. The goal of this analysis is to provide a tool for determining, on a financial basis, which option makes the most sense for you. To fairly evaluate the options, you need to consider the issues raised throughout this book that will determine how you can change from halocarbon refrigerants to other options, and in what timeframe you should do so. The approach presented in this chapter is only one of the many ways of evaluating these issues.

## REFERENCES

1. American Standard, Inc. *CFC Update,* Vol 16, November 1993, Arlington, VA.

2. Carrier "Comparing HCFC and HFC Refrigerants: A Primer for Application of Alternative Refrigerants," Communication white paper prepared by Jim Parsnow, Carrier Corporation, Syracuse, NY, April 1993.

3. Carrier, *Comparing HCFC and HFC Alternatives: A Primer for Application of Alternative Refrigerants,* by Jim Parsnow, Carrier Corporation, Syracuse, NY.

---

**SIMPLE PAYBACK PERIOD**

*Retrofit Example:*

| | |
|---|---|
| Investment Cost | $169,000.00 |
| Annual Cost Existing Equipment | $403,200.00 |
| Annual Cost for Retrofit | $386,910.00 |
| Annual Cost Savings for Retrofit | $16,290.00 |
| Years to Payback | 10 |

*Replacement Example:*

| | |
|---|---|
| Investment Cost | $306,900.00 |
| Annual Cost Existing Equipment: | $403,200.00 |
| Annual Cost for Replacement Equipment | $314,700.00 |
| Annual Cost Savings | $88,500.00 |
| Years to Payback | 3 |

FIGURE 9.

# Worksheets for Determining Your Best Option
*Retrofit, Replacement, or Retain?*

---

*YOUR EXISTING EQUIPMENT INFORMATION*

Type of Equipment _____

Age of Equipment _____

Type of Refrigerant _____

Charge _____

COP or kW/t _____

Electricity Cost ($/kWh) _____

Gas Cost ($/Therm) _____

Estimated Annual Hours of Operation _____

Estimated Leakage Rate _____

### YOUR OWN RETROFIT EXAMPLE:

**Investment Cost: (single year)**

Equipment

    O-rings, seals gaskets

    High efficiency purges

    Pressurizing System

Mechanical Room/Ventilation

    Vapor Detector

    Purge Piping

Refrigerant (initial charge)

R-   Credit

**Operations and Maintenance Cost: (yearly)**

**Energy Cost: (yearly)**

    COP

    kW/t

**Total Cost First Year**

**Total Annual Cost Succeeding Years**

---

### YOUR OWN RETROFIT:  Cost Calculation

| | | |
|---|---|---|
| **Investment Cost: (single year)** | | A |
| **Operations and Maintenance Cost: (yearly)** | | B |
| **Energy Cost: (yearly)** | | C |
| **Total Cost First Year** | | A+B+C |
| **Total Annual Cost Succeeding Years** | | B+C |

| | | |
|---|---|---|
| Year 1: | | $(1/(1.07))*(A+B+C)$ |
| Year 2: | | $(1/1.07)^2*(B+C)$ |
| Year 3: | | $(1/(1.07)^3)*(B+C)$ |
| Year 4: | | $(1/(1.07)^4)*(B+C)$ |
| | | |
| Year n: | | |

**Total  Cost Over n Years**

**YOUR OWN REPLACEMENT EXAMPLE:**

*Investment Cost: (single year)*
Equipment
Mechanical Room/Ventilation
Refrigerant (initial charge)
R-    Credit
*Operations and Maintenance Cost: (yearly)*
*Energy Cost: (yearly)*
   COP
   kW/t
*Total Cost First Year*
*Total Cost Succeeding Years*

---

**YOUR REPLACEMENT EXAMPLE:** *Cost Calculation*

| | | |
|---|---|---|
| *Investment Cost: (single year)* | | A |
| *Operations and Maintenance Cost: (yearly)* | | B |
| *Energy Cost: (yearly)* | | C |
| *Total Cost First Year* | | A+B+C |
| *Total Annual Cost Succeeding Years* | | B+C |

| | | |
|---|---|---|
| Year 1: | | $(1/(1.07))*(A+B+C)$ |
| Year 2: | | $(1/1.07)^2)*(B+C)$ |
| Year 3: | | $(1/(1.07)^3)*(B+C)$ |
| Year4: | | $(1/(1.07)^4)*(B+C)$ |
| | | |
| Year n: | | |

*Total  Cost over n years*

105

*YOUR OWN "RETAIN" EXAMPLE:*

*Investment Cost: (single year)*

Equipment

    O-rings, seals gaskets

    High efficiency purges

    Pressurizing System

Mechanical Room/Ventilation

    Vapor Detector

    Purge Piping

Refrigerant (initial charge)

*Operations and Maintenance Cost: (yearly)*

*Energy Cost: (yearly)*

    COP

    kW/t

*Total Cost First Year*

*Total Cost Succeeding Years*

---

*YOUR OWN "RETAIN" EXAMPLE: Cost Calculation*

| | | |
|---|---|---|
| *Investment Cost: (single year)* | | A |
| *Operations and Maintenance Cost: (yearly)* | | B |
| *Energy Cost: (yearly)* | | C |
| *Total Cost First Year* | | A+B+C |
| *Total Annual Cost Succeeding Years* | | B+C |
| *Total Cost Last Year* | | |

| | | |
|---|---|---|
| Year 1: | | $(1/(1.07))*(A+B+C)$ |
| Year 2: | | $(1/1.07)^2*(B+C)$ |
| Year 3: | | $(1/(1.07)^3)*(B+C)$ |
| Year 4: | | $(1/(1.07)^4)*(B+C)$ |
| | | |
| Year n: | | |

*Total Cost Over n Years*

## SIMPLE PAYBACK PERIOD for YOUR EQUIPMENT

### Retrofit Example:

Investment Cost      _____

Annual Cost Existing Equipment      _____

Annual Cost for Retrofit      _____

Annual Cost Savings for Retrofit      _____

Years to Payback      _____

### Replacement Example:

Investment Cost      _____

Annual Cost Existing Equipment:      _____

Annual Cost for Replacement Equipment      _____

Annual Cost Savings      _____

Years to Payback      _____

# CHAPTER 16

# CONCLUSIONS

The discovery of stratospheric ozone depletion, and the decision to discontinue production of halocarbon refrigerants, can be viewed as a roadblock and a headache—or as a turn in the road that will lead to unforeseen economic opportunities. We hope this book helps you to move down the road to opportunities. The key to confronting the future is arming yourself with the knowledge you require to make informed decisions, and ensuring that you have planned before you act. The technology that exists today, and that being developed for tomorrow, is more energy efficient, more specialized, and more diverse in providing opportunities for moving away from halocarbon refrigerants.

As this book has shown, decision makers must address a number of issues when making decisions about how to deal with their existing equipment. Among these are the need to take into account existing and possible future regulatory requirements for refrigerants, availability of chemicals and equipment, cooling load requirements and opportunities for responding to changes in cooling load, and energy efficiency and demand side management rebates that can provide incentives for equipment retrofit or replacement. But the single most important requirement facing any manager is to develop a plan.

This book has provided an overview of many of the issues to be considered. There are, of course, always more issues that could have been addressed. What we hoped to provide here is the starting block—the initial development of a plan and the evaluation of some options that can get you started thinking about how to approach the eventual phaseout of your own equipment. Once you have this initial information in hand, there are several places you can go for help and information. Equipment and refrigerant manufacturers are excellent sources of information. As more and more equipment users undertake equipment retrofits or replacements, the wealth of knowledge on how to approach these activities increases. You should plan to discuss your ideas and options with a manufacturer's representative to make sure that you are considering all the possible ways that your retrofit or replacement can be achieved.

While the absolute "best solution" to the question of how to phaseout halocarbon refrigerants may not exist today, the production phaseout effort has encouraged a reevaluation of the old technologies and provided an incentive for refinement of these technologies and development of new ones to meet current needs. Some exciting work is being done in the area of refrigeration research and development. New approaches to old problems are being undertaken to increase efficiency, capacity and performance of refrigeration and air conditioning equipment. Prompted by environmental concerns over ozone-depletion and global warming, emissions are being reduced and alternative refrigerants are being investigated.

What we hope to have shown in this book is that cost-effective approaches to refrigerant management are possible. The key is to begin planning, work with equipment and refrigerant manufacturers to obtain the best options for your needs, and look at other aspects of your building or operation that can help you conserve energy and lower costs. The optimum solution to phasing out your halocarbon refrigerants will take a systematic and organized effort that factors in your current and future situations and needs. Each solution will depend on many issues including corporate policy, personal preferences, and geographic location. One fact remains however, to do nothing is very much like being the "deer in the headlights" because right now this car is picking up speed and it has no brakes!

Good luck in your endeavors. We hope this book helps.

# Section V

# BIBLIOGRAPHY

# GLOSSARY

# INDEX

# BIBLIOGRAPHY

40 CFR 82. May 14, 1993. U.S. Environmental Protection Agency. "Protection of Stratospheric Ozone; Refrigerant Recycling; Final Rule." U.S. Code of Federal Regulations.

40 CFR 9 and 82. March 18, 1994. U.S. Environmental Protection Agency. "Protection of Stratospheric Ozone; Final Rule." U.S. Code of Federal Regulations.

*The Air Conditioning, Heating, and Refrigeration News.* August 16, 1993. "Score to Date: EPA Blows Whistle on 28 Alleged Venters; More Soon." Vol. 189, No. 16. Business News Publishing Co., Troy, Michigan.

*The Air Conditioning, Heating, and Refrigeration News.* December 6,1993. "What Desiccant-Based Cooling Is, How It Works." Vol. 190, No. 14. Business News Publishing, Co., Troy, Michigan.

*The Air Conditioning, Heating, and Refrigeration News.* December 27, 1993. "EPA Asks DuPont to Produce More CFC Refrigerants in 95." Vol. 190, No. 17. Business News Publishing Co., Troy, Michigan.

*The Air Conditioning, Heating, and Refrigeration News.* April 18, 1994. "Desiccant Compatibility with Alternatives." Vol. 191, No. 16. Business News Publishing Co., Troy, Michigan.

*The Air Conditioning, Heating, and Refrigeration News.* April 25, 1994. "Leak-Detection Plan Developed for Contractors to Increase Profits." Vol. 191, No. 17. Business News Publishing Co., Troy, Michigan.

*The Air Conditioning, Heating, and Refrigeration News.* June 20, 1994. "Replacement Compressors, Other Changes Solve Problems of Building's Aging Chillers." Vol. 192, No. 8. Business News Publishing Co., Troy, Michigan.

*Alternatives.* Spring, 1994. "ASHRAE Recommends ICI's KLEA 60 for Designation as Refrigerant 407A." ICI Klea. Wilmington, Delaware.

American Society of Heating, Refrigerating, and Air Conditioning Engineers. 1981.

*ASHRAE Handbook—1981 Fundamentals.* Atlanta, GA.

American Society of Heating, Refrigerating, and Air Conditioning Engineers. February 15, 1990. "Reducing Emission of Fully Halogenated Chlorofluorocarbon (CFC) Refrigerants in Refrigeration and Air Conditioning Equipment and Applications." *ASHRAE Guideline 3—1990.* Atlanta, GA.

American Society of Heating, Refrigerating, and Air Conditioning Engineers. 1991. *ASHRAE Handbook—1991 HVAC Applications.* I-P Ed. Atlanta, GA.

American Society of Heating, Refrigerating, and Air Conditioning Engineers. 1992. "Number Designation and Safety Classification of Refrigerants." *ASHRAE Standard.* ANSI/ASHRAE 34—1992. Atlanta, GA.

American Society of Heating, Refrigerating, and Air Conditioning Engineers. January 30, 1992. "Reducing Emission of Fully Halogenated Chlorofluorocarbon (CFC) Refrigerants in Refrigeration and Air Conditioning Equipment and Applications." *ASHRAE Guideline 3a—1992.* Atlanta, GA.

American Society of Heating, Refrigerating, and Air Conditioning Engineers. 1992. *ASHRAE Handbook—HVAC Systems and Equipment.* I-P Ed. Atlanta, GA.

American Society of Heating, Refrigerating, and Air Conditioning Engineers. 1993. *ASHRAE handbook—1993 Fundamentals.* I-P Ed. Atlanta, GA.

American Society of Heating, Refrigerating, and Air Conditioning Engineers. 1993. "Safety Code for Mechanical Refrigeration." *ASHRAE Standard—Public Review Draft.* BSR/ASHRAE 15-1992R. Atlanta, GA.

American Society of Heating, Refrigerating, and Air Conditioning Engineers. 1994. *ASHRAE Handbook—Refrigeration Systems and Applications.* I-P Ed. Atlanta, GA.

Barnett, C. J. May 16, 1994. "'How I spent last summer vacation'—HVAC professor learns CFC-ozone theory firsthand." *The Air Conditioning, Heating, and Refrigeration News.* Vol. 192, No. 3. Business News Publishing Co., Troy, Michigan.

Bas, E. July 12, 1993. "Add Refrigerant to Your List of Services." *The Air Conditioning, Heating, and Refrigeration News.* Vol. 189, No. 11. Business News Publishing Co., Troy, Michigan.

Bitzer Kühlmaschinenbau. December 1993. *Refrigerant Report 2: The Future Has Already Begun.* No. 9306 E, D-71044 Sindelfingen, Germany.

Bothwell, M., D. M. J. Sherbot, and C. M. Pollock. July 1, 1994. "Ecosystem Response to Solar Ultraviolet-B Radiation: Influence of Trophic-Level Interactions." *Science.* Vol. 265, No. 5168.

Brock, W.J. October 11, 1993. "When is a Refrigerant Really Safe?" *The Air Conditioning, Heating, and Refrigeration News.* Vol. 190, No. 6. Business News Publishing Co., Troy, Michigan.

Brumbaugh, J. E. 1987. *Heating, Ventilating, and Air Conditioning Library.* Vol. 3. Macmillan Publishing Co., New York.

Building Owners and Managers Association. 1993. *The Refrigerant Manual—Managing the Phase-Out of CFCs*. Washington, D.C.

Carrier Corporation. *Centrifugal and Rotary Screw Competitive Comparison Sheet*.

Carrier Corporation. *Choosing the Right Refrigerant for Tomorrow is Difficult: Information Sheet*. Carrier Corporation, Syracuse, New York.

Carrier Corporation. 1993. *Handbook for Responsible Refrigerant Management*. 1st Ed.

*Clean Air Report (CAR)*. December 17, 1992. "Antarctic Ozone Depletion Worse than Ever, U.S. Researchers Say." Inside Washington Publishers. Washington, D.C.

Culotta, E. July 1, 1994. "UV-B Effects: Bad for Insect Larvae Means Good for Algae." *Science*. Vol. 265, No. 5168.

Davis, L. May 2, 1994. "Don't Lose Your Cool Over Refrigerant Leaks." *The Air Conditioning, Heating, and Refrigeration News*. Vol. 192 , No. 1. Business News Publishing Co., Troy, Michigan.

DuPont. November, 1993. "Retrofit Guidelines for Suva HP62 in Stationary Equipment," "Cold, Hard Facts About Suva MP39, MP66, HP80, HP81, and HP62." *DuPont Technical Information Sheets*. DuPont, Wilmington, Delaware.

DuPont. December, 1993. "Retrofit Guidelines for Suva HP80," "Retrofit Guidelines for Suva MP39 and Suva MP66." *DuPont Technical Information Sheets*. DuPont, Wilmington, Delaware.

DuPont. January, 1994. "Suva 123 (Suva Centri-LP, HCFC-123) in Chillers," "Suva 134a (Suva Cold MP, HFC-134a) in Chillers," "Safety of Suva Refrigerants." *DuPont Technical Information Sheets*. DuPont, Wilmington, Delaware.

DuPont. April, 1994. "Retrofit Guidelines for Suva 134a (Suva Cold MP) in Stationary Equipment." *DuPont Technical Information Sheets*. DuPont, Wilmington, Delaware.

*Engineered Systems*. November/December, 1992. "Tech R-108: Take Action Today." Vol. 9, No. 11. Business news Publishing Co., Troy, Michigan.

*Engineered Systems*. June, 1994. Vol. 11, No. 6. Business News Publishing Co., Troy, Michigan.

*Engineered Systems*. July, 1994. "Thermal Storage By Design." Vol. 11, No. 7. Business News Publishing Co., Troy, Michigan.

Fields, S., and R. Flanagan. March 1994. "Wearing Thin—Ozone loss in the Northern Hemisphere has been getting worse. Why?" *Earth*. Vol 3, No 2. Kalmbach Publishing Co., Wisconsin.

Fisher, S. K., et al., Oak Ridge National Laboratory and C. L. Kusik, et al., Arthur D. Little, Inc. December 1991. *Energy and Global Warming Impacts of CFC Alternative Technologies*. Alternative Fluorocarbons Environmental Acceptability Study and U.S. Department of Energy. Oak Ridge, Tennessee.

General Services Administration. June 15, 1993. "Implementation of the National Capital Region's CFC Policy." *Memorandum*.

Gushee, D. E. 1993. "CFC Phaseout: Future Problem for Air Conditioning Equipment?" *CRS Report for Congress*. Washington, D.C.

ICI Americas Inc. October, 1993. *Data Sheets for KLEA Refrigerants*. ICI Klea. Wilmington, Delaware.

International Institute of Ammonia Refrigeration (IIAR). 1990. Proceeding of the *12th Annual Convention, International Institute of Ammonia Refrigeration*. March 4—7, 1990. Memphis, Tennessee.

International Institute of Ammonia Refrigeration (IIAR). 1985. *Industry in the Cold*. Chicago, Illinois.

Jazwin, R. 1994. *Technician's Guide to Certification*. Business News Publishing Co., Troy, Michigan.

Kleppe, J.S., and J & N Associates, Inc. December 27, 1993. "Leak Detection Depends on Tools, Techniques—Not 'Black Magic.'" *The Air Conditioning, Heating, and Refrigeration News*. Vol. 190, No. 17. Business News Publishing Co., Troy, Michigan.

Lewis, P.J, ed. 1993. *Proceedings of the 1993 Non-Fluorocarbon Refrigeration and Air-Condi-*

*tioning Technology Workshop.* June 23–25, 1993. Breckenridge, Colorado.

Mahoney, T.A. April 12, 1993. "Slow Conversion to Non-CFCs Worries Chiller Manufacturers." *The Air Conditioning, Heating, and Refrigeration News.* Vol. 188, No. 15. Business News Publishing Co., Troy, Michigan.

Mahoney, T.A. May 3, 1993. "A Contractor's Guide: Complying with EPA's Refrigerant Recycling Rule." *The Air Conditioning, Heating, and Refrigeration News.* Vol. 189, No. 1. Business News Publishing Co., Troy, Michigan.

Mahoney, T.A. April 11, 1994. "Statistical Panorama—Refrigerants." *The Air Conditioning, Heating, and Refrigeration News.* Vol. 191, No. 15. Business News Publishing Co., Troy, Michigan.

Mahoney, T.A. May 2, 1994. "EPA Sees 'Very Tight Supplies' of CFCs as Phaseout Approaches." *The Air Conditioning, Heating, and Refrigeration News.* Vol. 192, No. 1. Business News Publishing Co., Troy, Michigan.

McCoy, G.A., T. Litman, and J.G. Douglass. February, 1992. *Energy Efficient Electric Motor Selection Handbook.* Bonneville Power Administration, U.S. Department of Energy. Washington, D.C.

Morrisette, P.M. and N.J. Rosenberg. Winter, 1992. "Climate Variability and Development." *Resources.* No. 106. Resources for the Future, Washington, D.C.

Parker, G. August 26, 1993. *FY 1994 Hanford Energy Management (HEMC) Survey Study Request.* Energy Programs, Hanford Washington DOE-RL, Hanford, Washington.

Parsnow, J. January 11, 1993. *Comparing HCFC and HFC Refrigerants: A Primer for Application of Alternative Refrigerants.* Carrier Corp., Syracuse, New York

Parsnow, J. October, 1993. *Strategic Refrigerant Planning: A Step-by-Step Approach to Refrigerant Decision Making for Chiller Owners.* Carrier Corporation, Syracuse, New York.

Randazzo, M. June, 1994. "Esco Arranges utility Rebate, Loan to Fund Church's Retrofits." *Energy User News.*

*Report to the Secretary of Energy on Ozone Depleting Substances.* October 1989. "An Analysis of the Energy and economic Effects of Phasing Out Certain Organic Chlorine and Bromine Products."

Roose, R.W. 1978. *Handbook of Energy Conservation for Mechanical Systems in Buildings.* Van Nostrand Reinhold Co., New York.

Ruegg, R.T. November, 1987. *NBS Handbook 135: Life Cycle Costing manual for the Federal Energy Management Program.* National Bureau of Standards, US Government Printing Office, Washington, D.C.

Salas, C.E., and M. Salas. 1992. *Guide to Refrigerant CFCs.* Fairmont Press Inc., Lilburn, Georgia.

Salvato, J. *Environmental Engineering and Sanitation.* 4th Ed. Wiley/Interscience, John Wiley and Sons; NY, NY 1994

Siebert, B. April, 1993. "How to Convert CFC-11 Chillers to HCFC-123." *Heating/Piping/Air Conditioning.* Penton Publishing, Inc.

Smithheart, E.C. 1993. "Choosing A Building Chiller." In *Proceedings of the International CFC Conference.* pp. 250—258. October 20—22, 1993, Washington, D.C.

Sontz, S. and P. Tashian. July 18, 1994. "Listen: Do You Hear a Refrigerant Leak?" *The Air Conditioning, Heating, and Refrigeration News.* Vol. 192, No. 12. Business News Publishing Co., Troy, Michigan.

Swatkowski, L. June 13, 1994. *Perspective: CFC Phaseout and Global Climate Change.* Presented at the Environmental and Energy Study Institute Briefing. Association of Home Appliance Manufacturers.

Trane. November, 1993. *CFC Update.* Vol. 16. The Trane Co., Arlington, Virginia.

Trane. Spring, 1994. "CFCs, IAQ, and Energy: Cloudy Issues, Silver Linings." *Fresh Air: Exploring the Issues to Help You Build an Atmosphere of Success*. Vol. 2, No. 1. Trane Co., Marketing Communications Dept., LaCrosse, Wisconsin.

Trane. April 7, 1994. *New Energy Efficient "Earth-Wise" Trane Centrifugal Chiller*. The Trane Co., Publicity and Publications, Arlington, Virginia.

Trane. August 2, 1994. *Trane Engineered Conversion Performance Samples Sheet*. The Trane Co., LaCrosse, Wisconsin.

Trane. June, 1994. "Technical Information Brochure: High Efficiency Purge Systems." *Trane CFC Toolbox*. The Trane Co., Arlington, Virginia.

Trane. 1994. "Revisited: The Search for Chiller Efficiency." *Engineers Newsletter*. Vol. 23, No. 1. The Trane Co., Arlington, Virginia.

Turner, W.C. 1993. *Energy Management Handbook*. 2nd Ed. Fairmont Press, Inc. Lilburn, Georgia.

United Nations Environment Programme (UNEP) World Meteorological Organization. August 29, 1989. *Scientific Assessment of Stratospheric Ozone: 1989*. United Nations Environment Programme, Nairobi, Kenya.

United Nations Environment Programme (UNEP). 1989. *Action on Ozone*. Information and Public Affairs UNEP, Nairobi, Kenya.

United Nations Environment Programme (UNEP). May 1991. *Handbook for the Montreal Protocol on Substances that Deplete the Ozone Layer*. United Nations Environment Programme, Nairobi, Kenya.

United Nations Environment Programme (UNEP). 1991. *Environmental Effects of Ozone Depletion: 1991 Update*. UNEP Environmental Effects Panel Report, Nairobi, Kenya.

United Nations Environment Programme (UNEP). 1994. *Preliminary Draft 1995—Report of the Refrigeration, Air Conditioning, and Heat Pumps Technical Options Committee*. United Nations Environment Programme, Nairobi, Kenya.

United Nations Environment Programme (UNEP) Refrigeration, Air Conditioning, and Heat Pumps Technical Options Committee.

1991. "Refrigerant Data." *Montreal Protocol 1991 Assessment*. Chapter 2. United Nations Environment Programme, Nairobi, Kenya.

United Nations Environment Programme (UNEP) Refrigeration, Air Conditioning, and Heat Pumps Technical Options Committee. July, 1994. "Air Conditioning and Heat Pumps." *Montreal Protocol 1994/1995 Assessment*. Draft Chapter 7. United Nations Environment Programme, Nairobi, Kenya.

United Nations Environment Programme (UNEP) Refrigeration, Air Conditioning, and Heat Pumps Technical Options Committee. July, 1994. "Refrigerant Data." *Montreal Protocol 1994/1995 Assessment*. Draft Chapter 2. United Nations Environment Programme, Nairobi, Kenya.

United Nations Environment Programme (UNEP) Solvents Technical Options Committee. June 30, 1989. *Electronic, Degreasing, and Dry Cleaning Solvents Technical Options Report*. United Nations Environment Programme, Nairobi, Kenya.

United Nations Environment Programme (UNEP) Technology Review Panel. June 30, 1989. *Report of Technology Review Panel*. United Nations Environment Programme. Nairobi, Kenya.

United States Navy. December, 1993. *CFC—Halon News*. Vol. 3, No. 4. Navy CFC and Halon Clearinghouse. Arlington, Virginia.

United States Navy. June, 1994. "Tips and Tricks: Using CFC Replacements for Cascade Refrigeration Systems Retrofit." *CFC—Halon News*. Vol. 4, No. 2. Navy CFC and Halon Clearinghouse. Arlington, Virginia.

Vogelsberg, F.A. June 13, 1994. *CFC Phaseout and Global Climate Change: Challenges for Air Conditioning and Refrigeration*. Congressional Staff Briefing.

Weise, J.M. September 2, 1993. Personal Communication. The Trane Co., Arlington, Virginia.

York International. April, 1994. *CFC-Free Chiller Choices Sheet*.

# GLOSSARY

**absorbent**—A material that extracts one or more substances, for which it has an affinity, from a liquid or gas that changes physically, chemically, or both during the process.

**absorption**—Process whereby an absorbent extracts one or more substances present in an atmosphere or mixture that produces physical and/or chemical change in the absorbent.

**adsorbent**—Material with the ability to cause molecules of substances to adhere to its internal surfaces without changing the adsorbent physically or chemically.

**ambient air**—The air surrounding an object.

**atom**—The smallest individual structure which constitutes the basic unit of any chemical element. Consists of a complex arrangement of electrons revolving around a positively charged nucleus containing protons, neutrons, and other particles.

**azeotrope**—A liquid mixture that maintains a

constant boiling point and that produces a vapor of the same composition as the mixture itself.

**brine**—Any liquid cooled by the refrigerant and used for heat transmission without a change in state.

**BTU**—*British Thermal Unit*, the amount of heat required to raise the temperature of one pound of water from 59° F to 60° F.

**CFC**—*Chlorofluorocarbon*, a molecule composed of chlorine, fluorine, and carbon atoms is thought to contribute to ozone depletion.

**chemical stability**—The property of a chemical compound to not readily decompose and to not react with other compounds.

**compound**—A substance containing two or more elements chemically combined in fixed proportions in which the constituents of a compound lose their individual characteristics and create new characteristics when the compound is formed; a combination of two or more unlike atoms.

**compressor**—Machine that increases the pressure on a vapor by decreasing the volume of the vapor.

**condenser**—Part of a closed refrigeration system in which vapor discharged by the compressor is cooled by water or air, and is then returned to a liquid state.

**condensation**—The reduction of a gas to a liquid, typically as a result of a decrease in temperature.

**conduction**—Process of heat transfer through a material medium in which kinetic energy is transmitted from particle to particle by the medium's particles without the occurrence of large displacement.

**convection**—Heat transfer by the movement of fluid.

**cooling load**—Required rate of heat removal to maintain a constant temperature.

**COP**—*Coefficient of Performance*, = QL/W where QL represents the heat transformed from cold regions and W is the work required to drive the refrigerator.

**critical pressure**—The pressure under which a particular refrigerant vaporizes.

**critical temperature**—The temperature above

which a given gas cannot be liquefied, regardless of the pressure applied.

**de minimis releases**—Unavoidable releases which EPA requires be kept to a minimum during service, maintenance, or disposal of appliances.

**desiccant**—Any absorbent or adsorbent, liquid or solid, that removes water or water vapor from a material.

**evaporator**—Cooling system of refrigeration whereby the liquid refrigerant absorbs heat and is boiled into a vapor that carries off the absorbed heat.

**global warming**—Possible increase in atmospheric temperature due to the greenhouse effect.

**grandfather**—Provision in new law or regulation exempting those already in or part of an existing system being newly regulated.

**greenhouse effect**—The trapping of heat from sunlight at the earth's surface caused by atmospheric carbon dioxide and other greenhouse gases.

**greenhouse gases**—Gases, primarily $CO_2$, which facilitate the retention of heat from sunlight in the earth's atmosphere.

**GWP**—*Global warming potential*, relative index based on the global warming potential of $CO_2$ and calculated by the amount of gas remaining in the atmosphere and its ability to absorb heat over a 100-year lifetime.

**halocarbons**—A group of man-made chemicals formed by combining carbon, hydrogen, and chlorine, fluorine, or bromine, which are known as the halogens.

**HCFC**—*Hydrochlorofluorocarbon*, a molecule composed of hydrogen, chlorine, fluorine, and carbon.

**heat pump**—A refrigeration system designed to use alternately the heat extracted at a low temperature and the heat rejected at a higher temperature for cooling functions.

**heat sink**—High-temperature area to which thermal energy is transferred from a low-temperature area.

**hermetically sealed system**—System that is completely sealed by fusion, soldering, etc. to prevent the escape or introduction of a gas.

**Hg vacuum**—Also known as a mercury (Hg) barometer, an instrument that determines atmospheric pressure by measuring the height of a column of mercury that the atmosphere will actually support; the mercury is in a glass tube closed at one end and placed, open end down, in a well of mercury.

**high side**—Part of a refrigeration system that is operated at high, or condensing, pressure.

**hydrohalocarbon**—A halocarbon containing hydrogen, classified as an HCFC.

**inorganic compounds**—Compounds that are not classified as organic, most of which do not contain carbon and are derived from mineral sources.

**Intertropical Convergence Zone**—Area over the equator where air currents from the Tropic of Cancer and the Tropic of Capricorn converge and are driven upward, causing a continuous circulation of air in the northern and southern hemispheres.

**ion**—An atom or molecule that has acquired a net electrical charge either through the loss or gain of one or more electrons.

**latent heat**—The amount of heat required to effect a change of state at the boiling or freezing points of a substance.

**low-pressure refrigerant**—A refrigerant with a boiling point of around 75° to 80° F degrees which is in liquid form a room temperature.

**low-pressure system**—System where design pressure on the high side is less than approximately 15 psia.

**low side**—Part of a refrigeration system that is operated at low, or evaporating, pressure.

**lubricant**—A substance for reducing friction by providing a smooth film as a covering over parts that move against each other.

**miscible**—Capable of being mixed.

**molecule**—The smallest particle of an element or compound that can exist in the free state and still retain the characteristics of the element or compound; a combination of two or more atoms.

**organic compounds**—Most carbon-containing compounds; chemical compounds based on carbon chains or rings and also containing hydro-gen with or without oxygen, nitrogen, and other elements.

**ODP**—*Ozone depletion potential*, the potential of a substance containing chlorine to react with ozone and the degree to which such reactions could take place.

**ozone layer**—Also known as the ozonosphere, the atmospheric layer extending from 6 mi. to 30 mi. above the earth's surface in which there is an appreciable concentration of ozone, which absorbs much of the ultraviolet radiation from the sun and prevents some heat loss from the earth.

**photosynthesis**—The production of organic compounds from carbon dioxide and water. Occurs in green plant cells by the interaction of chlorophyll and light.

**positive displacement system**—A compressor that confines successive volumes of fluid within a closed space in which the pressure of the fluid is increased as the volume of the closed space is decreased.

**prime mover**—The component of a system that transforms energy from the thermal or the pressure form into the mechanical form.

**purger**—A device that removes non-condensable gas from refrigerant condensers, low-concentration liquid from absorption system evaporators, or air from hot water or steam systems.

**reclamation**—The recovery of useful materials from waste products, reprocessing of refrigerants to meet the purity requirements of the ARI-700 standard and verifying purity through testing at EPA-certified labs.

**recovery**—The removal of valuable materials from waste products; removal of refrigerant from an appliance by specifically designed equipment, with the refrigerant not necessarily tested or processed to ensure that it is free from contaminants.

**recycling**—The reuse of a refrigerant that has been recovered and purified; the extraction and cleaning of a refrigerant for reuse.

**refrigerant**—Fluid used for heat transfer in a refrigeration system that absorbs heat at a low temperature and low pressure and rejects heat at a high temperature and high pressure.

**relative humidity**—The ratio of the water vapor present in the air to the water vapor

121

present in saturated air at the same barometric pressure and temperature.

**remaining useful life**—The estimated amount of time that current equipment can continue to be operated based on an average lifetime for commercial and industrial process equipment of 25 to 30 years.

**saturated compound**—An organic compound containing no double or triple bonds and having no free valence electrons.

**sensible heat**—Heat associated with an immediate change in temperature.

**solubility**—The capability of being dissolved, the amount of a substance that can be dissolved in a given solvent under specific conditions.

**solution**—A single, homogenous liquid, solid, or gas that is a mixture in which the components are uniformly distributed.

**solvent**—A substance that dissolves a solute.

**specific humidity**—The mass of water vapor to the mass of dry air contained in the sample.

**stratosphere**—The atmospheric zone above the troposphere, extending from approximately 8 to 30 miles above the earth's surface, in which the temperature is fairly constant.

**temperature**—Thermal state of matter in reference to its tendency to communicate heat to matter in contact with it.

**thermal currents**—Warm air currents that move as a result of the intermingling of cold air currents.

**thermodynamic**—Pertaining to heat energy and its transformation to and from other forms of energy.

**thermophysical properties**—Qualities that help determine the suitability as a refrigerant and indicates the potential efficiency and economy of the refrigeration cycle.

**troposphere**—Atmospheric zone that extends from the earth's surface to the stratosphere; in which stratum clouds form, convective disturbances occur, and the temperature typically decreases with altitude.

**ultraviolet radiation**—Radiation given off by the sun; lies just beyond the violet end of the visible spectrum and has a wavelength shorter than 400 nanometers (nm).

**UV-B**—Type of ultraviolet radiation with a wavelength of 280nm—320 nm from which the ozone layer offers protection; the energy in UV-B wavelengths tends to be very biologically effective and has been shown to produce serious health disorders, such as skin cancer, and to be a mutagen in both animal and bacterial cells.

**vapor**—A gas near to equilibrium with the liquid phase of the substance and that does not follow the gas laws.

# INDEX

# STRATEGIES FOR MANAGING OZONE-DEPLETING REFRIGERANTS
## Confronting the Future

Katharine B. Miller, Charles W. Purcell, Jennifer M. Matchett, and Marjut H. Turner

As a result of the *Montreal Protocol on Substances that Deplete the Ozone-Layer*, and the Clean Air Act Amendments of 1990, the production of most ozone-depleting substances will be discontinued at the end of 1995. Users of these substances will need to develop strategies for retrofitting or replacing refrigeration and air-conditioning equipment in a cost-effective manner. The purpose of this book is to provide a comprehensive strategy for the phaseout of equipment that uses CFC-refrigerants. The book provides the reader with essential information on alternatives to ozone-depleting refrigerants, critical data that must be gathered to make informed decisions regarding equipment retrofit or replacement, and an implementable approach for arriving at cost-effective solutions. It is the first book to provide an implementation methodology that combines consideration of the economic, environmental, and energy components of the decision making process.

The book includes worksheets and example forms that will be immediately useful in refrigerant management activities. It also includes answers to the most frequently asked questions on how refrigerant-CFC users can meet the requirements of the current regulations–and stay in business.

> "*Strategies for Managing Ozone-Depleting Refrigerants: Confronting the Future* provides a strategic business approach to managing ozone-depleting refrigerants...[it] will be useful to environmental managers, building/facility managers, supervisors, and anyone else with an interest in refrigerant management."
> —**Dr. Stephen O. Anderson**, Co-Chair Technology and Economic Assessment Panel, United Nations Environment Programme

**About the Authors:** The authors, members of Battelle's Technology Planning and Analysis Center, have worked extensively with the Environmental Protection Agency, Department of Defense, Department of Energy, United Nations Environment Programme, and the North Atlantic Treaty Organization in promoting cost-effective alternatives and substitutes to ozone-depleting substances. They have been instrumental in the development of CFC management plans for both large and small organizations.

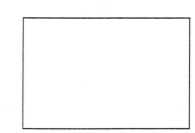